HEXHAM
LOCAL HISTORY
SOCIETY

This facsimile edition, published in 2015, of Robert Rawlinson's report into the Sanitary Conditions of the People of Hexham is the result of a collaboration between the Society of Antiquaries of Newcastle upon Tyne and Hexham Local History Society and is published in paperback and digital form by Hexham Local History Society.

PRICE £7.50

HEXHAM
LOCAL HISTORY
SOCIETY

Rawlinson's report is extremely rare with only a few copies mentioned in national libraries. The Society wishes to thank Denis Peel, Honorary Librarian of the Society of Antiquities at Newcastle upon Tyne, and a member of this Society, for facilitating the copying of the report. Its fragile condition introduced some restrictions on the photography which was converted into a digital format suitable for print.

Photography: Patrick Lindsay and Jim Hedley
Design & Artwork: Peter Rodger

Front cover image: 'Deaths Dispensary'
Source: The London Museum of Water & Steam, TW8 0EN

© Hexham Local History Society, 2015
All rights reserved
www.hexhamhistorian.org

ISBN 978 0 956 5078 6 0
Occasional Publication No.13

Printed by Lightning Source UK, Milton Keynes, MK11 3LW

Introduction

In 1854, Hexham made its first proper steps into the modern world. Until that date, there had been few forays into modernity as we know it. The coming of the railway was without doubt of great importance to the town in its relationship with the outside world, particularly the eastern portion of the Tyne Valley. But, in a sense, the railway was thrust upon the town, though money from it was used for the building of the Subscription School. The most significant change generated within the town before 1854 was the building of the gas works, which meant that the town was better lit than hitherto with its flickering oil lamps.

The decision to build the gas works was made by the Vestry. That body and the Borough Court, which were part of the 'ancient government of the town', were the major administrative forces shaping the development of Hexham. That was to change following the first Public Health Act in 1848 with the establishment of the General Board of Health with members appointed by the Crown which was to generate sanitary improvement in the country. The move to establish a Local Board in Hexham began when 152 of the rated inhabitants petitioned the General Board to that effect on 30 August 1851. Just over a year later, in September 1853, Robert Rawlinson (1810-1898), a Superintending Inspector of the General Board visited the town to examine its sanitary state and hear from leading citizens their views on related issues. Rawlinson's Report was produced as a result of the visit. The text leaves one in no doubt as to the incredible insanitary state of the town. One is also left with the view that this state owed much to the failure of the antiquated, creaking, nature of the town's administration. Not surprisingly, though going against the increasing opposition to change within the town generated by Rawlinson's visit, it was reported in the 4 October issue of The London Gazette that the General Board applied the Public Health Act of 1848 'throughout the township of Hexham'. The elected Local Board was to consist of nine members. On 15 December 1853, their election took place. Around a decade on, the burgeoning nature of the democratic government within the town was strengthened by the arrival of two weekly papers – the Liberal Hexham Courant in 1864 and the Conservative Hexham Herald in 1866. Proper democratic involvement of at least part of the populace in the running of the town had arrived.

Within the pages of Rawlinson's report the sanitary problems are described in considerable detail. But as well as this, the report touches upon many aspects of life in Hexham providing a fine picture of life in the town in 1853. For these reasons, it is very good that Rawlinson's Report is at last reaching a wider public.

What of Rawlinson? Coming to his post on the General Board, from a wide experience in civil engineering, his future career revolved almost wholly round sanitary issues. He was knighted in 1883 and became President of the Institute of Civil Engineers in 1884.

David Jennings June 2014

PUBLIC HEALTH ACT
(11 & 12 Vict. Cap. 63.)

REPORT

TO THE

GENERAL BOARD OF HEALTH

ON A

PRELIMINARY INQUIRY

INTO THE SEWERAGE, DRAINAGE, AND SUPPLY OF
WATER, AND THE SANITARY CONDITION
OF THE INHABITANTS

OF THE TOWN AND TOWNSHIP OF

HEXHAM,

IN THE PARISH OF HEXHAM, IN THE COUNTY OF
NORTHUMBERLAND.

By ROBERT RAWLINSON, Esq.

SUPERINTENDING INSPECTOR.

LONDON:
PRINTED BY GEORGE E. EYRE AND WILLIAM SPOTTISWOODE,
PRINTERS TO THE QUEEN'S MOST EXCELLENT MAJESTY.
FOR HER MAJESTY'S STATIONERY OFFICE.

(15)   1853.

NOTIFICATION.

THE General Board of Health hereby give notice, in terms of section 9 of the Public Health Act, that on or before the 21st day of August next, being a period of not less than one month from the date of the publication and deposit hereof, written statements may be forwarded to the Board with respect to any matter contained in or omitted from the accompanying Report on a preliminary Inquiry into the Sewerage, Drainage, and Supply of Water, and the Sanitary Condition of the Inhabitants of the Township of HEXHAM, in the County of Northumberland; or with respect to any amendment to be proposed therein.

By order of the Board,
C. MACAULAY, *Secretary*.

Whitehall, 14th June 1853.

CONTENTS.

	Page
INTRODUCTION	7
Notice of Inquiry	7
Proof that Notice was properly posted and advertised	8
ANCIENT HISTORY of HEXHAM	8
ANCIENT CHURCH	10
TOPOGRAPHY	10
LOCAL GEOLOGY	10
LOCAL METEOROLOGY	11
Fall of Rain at West Denton	12
Fall of Rain at Fallowfield	12
Fall of Rain at Whittle Dean	13
Remarks on do.	13
MODERN TOWN of HEXHAM	13
LOCAL GOVERNMENT	17
LOCAL TRADE	17
MARKETS and FAIRS	17
Hotels, &c.	18
Slaughterhouses	18
Baths	18
ROADS	18
RENTAL of LAND	18
INQUIRY	18
Names of Gentlemen present	18
Preliminary Remarks	19
EVIDENCE—	
Boundaries of Hexham Township	19
Area of Hexham Township	20
Population of Hexham Township	20
Sewerage and Drainage	20
Water Supply	20
Lighting	20
State of Burial Grounds	20
Number of Inhabitants	20
Annual Rateable Value of Property in the Township	21
Poor's Rate	21
Highway Rate	21
Lamp and Watch Rate	21
Church Rate	21
Sanitary Condition of the Inhabitants	21
Rateable Value of Hexham Town	21
Annual Rateable Value, according to Rate Book, June 1852	21

(15) A 2

CONTENTS.

Page

EVIDENCE of RESIDENT GENTLEMEN—
Mr. Thos. Jefferson, Surgeon - - - - - 22
Mr. Robert Stokoe, „ - - - - - 22
Mr. W. A. Temperley, Corn Merchant - - - - 22
Mr. W. Pearson, Surgeon - - - - - 23
Mr. Thos. Stainthorpe, Surgeon - - - - 23
Messrs. J. Nicholson and J. B. Maughan - - - - 23

LOCAL MORTALITY - - - - - - - 23
Mr. Stainthorpe's Statement - - - - - 23
Inflamatory, remittent, continued, and Typhus Cases - - 23
Cases of Small-pox - - - - - - 24
Small-pox Cases, arranged into Wards - - - - 24
Table of Fever Cases - - - - - - 25
Mortality in Hexham Township - - - - 26
Deaths in Hexham Township, exclusive of Workhouse - - 26
Average Deaths of Hexham Township - - - 26

POOR RELIEF - - - - - - - 27

INSPECTION of TOWN and DISTRICT - - - - 27
Names of Gentlemen who attended during the Inspection - - 27
Abstract of Notes taken - - - - - 27
Water Supply - - - - - - - 27
Draiange, &c. - - - - - - - 28
Condition of Houses, &c. - - - - - 28
The Seal—Recreation Ground - - - - - 29
Lighting and Watching - - - - - - 30
Statement of the Receipt and Expenditure of the Lamp and Watch
Inspectors for Seven Years - - - - - 31
General Statement of Income and Expenditure of Gas-works between
19th December 1849 and 18th December 1850 - - - 32
Remarks - - - - - - - - 33

BURIAL GROUNDS - - - - - - - 33
Report on Burial Grounds - - - - - 33
Burials in Roman Catholic Cemetery - - - - 34
Fees for Marriages, Burials, &c. in Hexham Parish Church - - 34

ABSTRACT of the REPORT and SPECIFICATION of WORKS REQUIRED - 35
Specification of Works required—Sewerage and Drainage - - 37
Waterworks - - - - - - - 37

ANALYSIS OF LOCAL WATERS - - - - - 38
Samples of Waters - - - - - - 38
Report on Samples of Waters, by Mr. Holland - - - 38
Remarks on Water Supply - - - - - 39

PROPOSED WORKS - - - - - - - 40
Estimate for Water Supply - - - - - 40
„ Sewerage and Drainage - - - - - 41
„ Abstract - - - - - - 41
„ Annual Expenses - - - - - 41
„ Income from Water - - - - - 41

ROAD FORMATION AND STREET PAVEMENT - - - - 42

CONTENTS.

ABSTRACT of PUBLIC HEALTH ACT, with REMARKS as to the POWERS it contains - - - - - - - 43

SUMMARY OF CONCLUSIONS AND RECOMMENDATIONS - - - 50
   Conclusions - - - - - - 50
   Recommendations - - - - - - 51

APPENDICES.

APPENDIX A.—Remarks, &c. - - - - - - 53
   Petition for Application of Act - - - - - 53
   Petition against the Act - - - - - - 54
   Memorial against the Inquiry - - - - - 56
   Water Supply—Ratepayers' Committee Report on - - 57
   Engineer's (Mr. Nicholson's) Report on Water Supply - - 60
   Remarks - - - - - - - - 62
   Resolutions at Moot-hall, 15th August 1851 - - - - 62
   Report of the Committee to a Public Meeting, 4th August 1851, as to the Water Supply - - - - - - 63

APPENDIX B.—Remarks, Local Reports, &c. - - - - 66
   Water Supply - - - - - - - 66
   Sewerage and Drainage - - - - - - 71
   Sanitary Report - - - - - - 73
   Burial Ground Report - - - - - 77
   Common Lodging-houses Report - - - - - 78
   Public Roads and Streets, Report on - - - - - 80
   Boundary Report - - - - - - 81

APPENDIX C.—Particulars as to the ancient Form of Government, by Jasper Gibson, Esq. - - - - - - 81
   Ancient Government, Division of Town, Oaths of Local Officers, &c. 81
   Borough Jury - - - - - - - 81
   Scavengers - - - - - - - 82
   Powers enforced - - - - - - - 83
   Constable's Oath - - - - - - 84
   The Oath of the Market-keepers, &c. - - - - 84
   The Common-keeper's Oath - - - - - 85
   The Surveyor's Oath - - - - - - 85
   The Oath of the Four-and-twenty - - - - - 85
   The Ale-taster's Oath - - - - - - 85

PUBLIC HEALTH ACT (11 & 12 Vict. c. 63.)

Report to the General Board of Health on a Preliminary Inquiry into the Sewerage, Drainage, and Supply of Water, and the Sanitary Condition of the Inhabitants of the Town and Township of HEXHAM, *in the Parish of Hexham, in the County of Northumberland,* By ROBERT RAWLINSON, Esq., *Superintending Inspector*

London, March 1853.

MY LORDS AND GENTLEMEN,

A PETITION (see page 53) from the township of Hexham, numbered 2,982, and signed by 152 persons, (being more than one tenth of the ratepayers), praying that an inquiry might be made, and that the Public Health Act might be applied to the township, was received by your Honourable Board on the 31st August 1851. Petitions and memorials against the inquiry were subsequently presented, the preambles of which are given in the Appendix.

Having received instructions to make inquiry in Hexham, the notice, as under, was advertised and published as the Act directs; and in the place named, and at the time as set forth, I held open court, made inquiry, received evidence, &c., and now beg respectfully to submit this as my Report.

"PUBLIC HEALTH ACT, 1848.—11 & 12 VICT. C. 63.

" NOTICE.—Whereas, in pursuance of the Public Health Act, 1848, the General Board of Health have directed Robert Rawlinson, Esquire, one of the Superintending Inspectors appointed for the purposes of the said Act, to visit the township of Hexham, in the county of Northumberland, and there to make public inquiry, and examine witnesses with respect to the matters following ; that is to say,—

" The sewerage, drainage, and supply of water ;
" The state of the burial-grounds ;
" The number and sanitary condition of the inhabitants ;
" The Local Acts of Parliament (if any) for paving, lighting, cleansing, watching, regulating, supplying with water, or improving, or having relation to the purposes of the said Act ;
" The natural drainage areas ;

" The existing parochial, or other local boundaries;
".The boundaries which may be most advantageously adopted for the purposes of the said Act ;
" And other matters in respect whereof the General Board of Health is desirous of being informed, for the purpose of enabling them to judge of the propriety of reporting to Her Majesty, or making a provisional order, with a view to the application of the said Act, or any part thereof, to the said township.

"Now, therefore, I, the said Robert Rawlinson, do hereby give notice, that on the twenty-ninth instant, at ten o'clock in the forenoon, at the Board of Guardians Room, I will proceed upon the said inquiry, and that I shall then and there be prepared to hear all persons desirous of being heard before me upon the subject of the said inquiry.

" Dated this eighth day of September 1852.
" ROBERT RAWLINSON."

NOTICE PROPERLY POSTED AND ADVERTISED.—Mr. *Ralph Errington Ridley* proved that the above notice had been affixed on the doors of the principal churches, chapels, buildings, and places where public notices are usually affixed within the township, namely, the Established Church, Methodist chapel, Independent chapel, Roman Catholic chapel, Primitive Methodist chapel, Presbyterian chapel, Secession chapel, &c.; and also that it had been duly advertised in the " Newcastle Courant," " Newcastle Chronicle," " Newcastle Journal," Newcastle Guardian," and " Gateshead Observer."

ANCIENT HISTORY.*—Hexham† is a town of great antiquity, antiquaries believing it to have been a Roman station.

In the year 674 Hexham was the see of a bishop, but the diocese was so harassed by the Danes, that no man would accept the bishopric, so that in the year 883 it was united to Lindisfarne. A monastery was founded here in the year 1112, with liberties that procured the district the name of a shire.

---

* Mr. Jasper Gibson has kindly corrected these few remarks on the ancient history of Hexham.

† *Hexham.* The town is situate on the banks of the brooks Hextol (Saxon, Height of source,) and Halgut (Halig-gut-stede, the dwelling by the holy stream). There is also East-burn, commonly called *Skinners-burn.* The name of the town, at early periods, was written Hutoldesham, Hestoldesham, Hextoldesham, contracted by the Normans into Hexham, Hexam, and in some old deeds, Exam. The brook Hextol is now called Cockshaw-burn, from the local name of the suburb through which it flows. The Halgut is, for a similar reason, Cowgarth-burn. The name Seal-burn is common to both streams, from the Seal or priory grounds which are situate between the two streams.

In the year 1296 the town and priory were burned by the Scots; and in the reign of Edward III., David king of Scots entered the borders by Liddel Castle, and pillaged the town. In the year 1461, on the plains called "the levels," the battle of Hexham was fought between the adherents of the houses of York and Lancaster.

The Act 27 Henry VIII. c. 24. intituled "An Act for re-continuing liberties in the Crown" (a general Act regulating or abridging the prerogatives and authorities exercised in counties palatine, &c.) had no special reference to Hexham, except in so much that there is a proviso that the Archbishop of York and his temporal chancellor for the time being of the shire and liberty of Hexam, otherwise called Hextoldsham, and every of them, should from thenceforth be justices of the peace within the said shire and liberty of Hexam, otherwise called Hextoldsham, &c.

The palatinate jurisdiction in Hexham and the adjacent district (much larger than what is now called Hexhamshire), dates from a very early period, and from the time of Henry I. to that of Henry VIII. was vested in the Archbishop of York. In the latter reign it was exchanged with and became vested in the Crown; and by statute 14 Elizabeth, c. 13., it was enacted, "that the territoree, franchise, and liberty of Hexam and Hexamshire, with the libertys of the same, may be, is, and shall be from henceforth taken to be within, and part, parcel, and member of the said county of Northumberland," &c.—See *Wright*, 108, 109.

Notwithstanding this Act, the court of pleas of the ancient county palatine, now called "the regality or manor of Hexham," still remains, and is now presided over by James Losh, Esq., under the title of the "learned steward of the same court." In this court actions to any amount, arising within the jurisdiction, may be tried. It is held twice a year, within a month after Easter and Michaelmas respectively, and it is usual to hold the ordinary manorial courts with it.

The court baron, for the recovery of small debts, and the customary court, for the transaction of matters relating to copyholds, are also held at other times, and are then presided over by the bailiff, under the title of "steward of the same court."

The manor of Anick Grainge, which is partly within the regality or manor of Hexham, formerly belonged to the priory of Hexham, and became vested in the Crown upon the dissolution of monasteries. It has the usual manorial courts.

Both the regality or manor of Hexham and the manor of Anick Grainge belong to Wentworth Blackett Beaumont, Esq., M.P.

ANCIENT CHURCH.—The former church is said to have been constructed by workmen brought from Italy, and is described to have been of "great beauty and elegance." No portion of this structure now remains. The present church is a mixture of Saxon and Gothic architecture; there are many ancient tombs in it. This church possessed the privilege of sanctuary, until taken away by Henry VIII. The privilege extended one mile from the church in four directions, the boundaries being marked by crosses. Excommunication, or heavy penalties, were levied with the utmost severity on whomsoever should dare to violate the sanctuary. The "stool of peace" is still preserved. At the west end of the church are remains of the former priory.

TOPOGRAPHY.—The township of Hexham is in the parish of Hexham, in the county of Northumberland. There are five townships in the parish of Hexham, namely, township of Hexham, Hexhamshire high quarter, Hexhamshire low quarter, Hexhamshire middle quarter, and Hexhamshire west quarter.

Hexham township is divided into four wards, namely, Market-street, Priest-popple, Hencotes, and Gilligate, each of which did appoint a separate constable until the adoption of the Lighting and Watching Act. Overseers, churchwardens, and surveyors are nominated at vestry meetings for each ward; but their appointments really take place for the whole township. The magistrates appointments are for the whole township, and the several parties appointed to office act for the township at large. Within the district there are the parish church and Whitley chapel. The burial-ground attached to Whitley chapel is used by high, low, and middle quarters. West quarter residents bury their dead in Hexham (the parish) churchyard.

LOCAL GEOLOGY.—Hexham stands upon millstone grit, the Newcastle coal-field cropping out to the east, and mountain or carboniferous limestone covering a large area of country to the north. Millstone grit consists of quartzose grits, with shales, coal, ironstone, &c., the regular strata being in most parts covered with alluvium, consisting of marl, clay, gravel, sand, and varying mixtures of these. Springs of pure water generally abound in the millstone grit; many such issue above the town, at an elevation sufficient to supply the whole of the inhabitants at high pressure. Nature has done much for Hexham to make it a place of healthy residence. Man has hitherto done nothing worthy of note to improve the site, or even to make available the natural advantages,—but the contrary; by surrounding dwelling-houses with his own refuse, he has generated fever and other zymotic diseases, until the annual mortality (considering the population) ranks Hexham amongst the most unhealthy towns in the kingdom. A favourable subsoil is corrupted by infiltration of human refuse, and pure water is rendered foul and impure by the faulty mode of collection. The wholesome air of the district is tainted by foul emanations from privies, cesspools, middens, slaughter-houses, and grave-yards; and yet there are medical gentlemen practising in the town, who, according to their evidence, think this state of things cannot be improved.

LOCAL METEOROLOGY.—The results of long continued meteorological observations are always valuable; the facts must, however, be legitimately applied. Averages of rainfall, for a number of years, are useful in making general comparisons, such as one district with another district; it is, for instance, useful to know that the average rain-fall on the east coast of England is about 20 inches, and on the west coast about 40 inches; as also, that the average for all England is about 36 inches. But such averages of any limited district are almost useless for practical purposes of water supply or drainage, for the following reasons:—The dimensions of sewers and drains must not be fixed to accommodate averages—they have to work efficiently during excesses, and, independent of both maximum and minimum falls of rain. The question of sewerage and drainage is, however, intimately connected with the character and formation of the urban and suburban contours and gradients. The natural surface must be studied before the dimensions of sewers and drains are decided upon, or excess of expenditure may be incurred. If local rain-floods have hitherto passed away over a surface, it surely will not be advisable to provide new channels below the levels of cellars to contain and pass off future storm waters. Sewers of the capacity necessary to contain storm waters must be, under any circumstances, in cost excessive, in many places impossible. With respect to water supply, averages of years are equally useless, the available quantity will be governed, in many cases, by the minimum, even where storage reservoirs are used, as a drought of three or four months duration may take place, and few impounding reservoirs provide for more than this period.

The following tables of local rain-fall may prove useful:—

FALL of RAIN as recorded at WEST DENTON, near Newcastle-upon-Tyne, and at the Royal Observatory, GREENWICH.

	1845.		1846.		1847.		1848.	
	West Denton.	Greenwich.	West Denton.	Greenwich.	West Denton.	Greenwich.	West Denton.	Greenwich.
Yearly amounts	37·88	22·4	40·26	25·3	29·85	17·8	40·36	30·2

FALL of RAIN as registered at FALLOWFIELD MINES, about 2 miles north of Hexham, for the year 1852.

Date.	Inches.	Date.	Inches.
3 January	0·50 ⎫		Brought forward 19·43
10 ,,	1·54 ⎪	3 July	1·03 ⎫
17 ,,	1·07 ⎬ 5·84	10 ,,	0·43 ⎪
24 ,,	1·26 ⎪	17 ,,	0·53 ⎬ 3·37
31 ,,	1·47 ⎭	31 ,,	0·00 ⎪
		24 ,,	1·38 ⎭
7 February	2·42 ⎫		
14 ,,	0·39 ⎪	7 August	1·07 ⎫
21 ,,	1·89 ⎬ 4·70	14 ,,	1·42 ⎪
28 ,,	0·00 ⎭	21 ,,	1·97 ⎬ 5·13
		28 ,,	0·67 ⎭
6 March	0·08 ⎫	4 September	0·17 ⎫
13 ,,	0·00 ⎪	11 ,,	0·96 ⎪
20 ,,	0·00 ⎬ 0·42	18 ,,	0·00 ⎬ 1·55
27 ,,	0·34 ⎭	25 ,,	0·42 ⎭
3 April	0·78 ⎫	2 October	4·45 ⎫
10 ,,	0·36 ⎪	9 ,,	1·63 ⎪
17 ,,	0·00 ⎬ 1·14	16 ,,	0·11 ⎬ 9·77
24 ,,	0·00 ⎭	23 ,,	0·31 ⎪
		30 ,,	3·27 ⎭
1 May	0·85 ⎫	6 November	1·66 ⎫
8 ,,	0·15 ⎪	13 ,,	1·62 ⎪
15 ,,	1·39 ⎬ 3·76	20 ,,	2·47 ⎬ 6·94
22 ,,	1·35 ⎪	27 ,,	1·19 ⎭
29 ,,	0·02 ⎭	4 December	1·46 ⎫
		11 ,,	0·83 ⎪
5 June	0·42 ⎫	18 ,,	4·19 ⎬ 9·95
12 ,,	1·54 ⎪	25 ,,	0·74 ⎭
19 ,,	0·89 ⎬ 3·57	1 January	2·73
26 ,,	0·72 ⎭		
Carried forward	19·43	Rain-fall for the year	56·14

Hexham Town.

FALL of RAIN according to a rain-gauge kept by the Newcastle Water Company at WHITTLE DEAN, about 11 miles west of Newcastle-upon-Tyne, and 12 miles east of Hexham.

Dates.	1850.	1851.	1852.
From 1st January to 30th June - -	7·20	10·40	11·70
From 1st July to 31st December - -	10·48	10·93	23·02
Yearly amounts - -	17·68	21·33	34·72

REMARKS.—This table shows the yearly fall of rain in 1852 to be nearly double in volume that of 1850; and the first half of the year 1852 is nearly one half less than the latter half of the same year; the daily differences were no doubt much greater. Sometimes, during a thunderstorm, more than the average of a month will fall within one hour.

MODERN TOWN.—Hexham is irregularly built. The modern town occupies its ancient site, and is traversed by three small streams or burns. It stands on ground sloping towards the river Tyne. The streets are irregular on plan, the houses are of unequal height, and have been built without order, symmetry, or regularity; there are fronts of stone and others of brick; some are plastered and rough-cast; a few of the old houses remain thatched, and many have the heavy grey flag-slates of the district. The houses are confined and cramped behind, the land for the most part sloping towards the street. There are pigsties, privies, cesspools, and foul middens crowded near the dwelling-houses and narrow tenements, and the slaughter-houses are said to be a great nuisance. The streets are partially formed with broken stone (Macadam); they are dirty in wet weather, and dusty in dry weather; some of the streets and side channels are paved with pebble boulders, and a portion of the street, where there is most business, is paved with square sets. The footwalks, within the town, are partly paved with small boulders and are partly flagged; many of the footwalks are in holes and out of repair. The town is lighted by gas. There are some isolated rubble sewers and drains, but none that can act efficiently for purposes of house drainage. Water is, to a limited extent, supplied in open troughs and pants; this is muddy during rain-fall, and more or less impure at all times. There are private pumps and wells.

In a recent history of the town it is stated,—
" The present state of the streets in Hexham is not conducive

to the health of the inhabitants ; they are generally narrow, and one large house is often tenanted by several poor families. The pavement has been laid with little regard to the comfort or convenience of pedestrians, and lamps are scarcely known.* The town is so situated that almost every street is formed on a descent, an advantage which might be made conducive to a state of greater cleanliness. The vicinity of the river, the general excellence of the roads, the bridge, and other advantages are favourable circumstances, of which the inhabitants do not sufficiently avail themselves."

With the exception of the establishment of gas-works and the improvement of street lighting, Hexham remains as described above. The streets are imperfectly paved; such sewers as have been made, are formed of dry rubble stone, and consequently only add to previously existing nuisances; refuse is retained for long periods behind the houses and in contact with house-walls, and is ultimately emptied out over the streets in the day-time, to the great incovenience of the inhabitants. All the natural advantages of this district in site, subsoil, and water are, for want of proper works and efficient regulations, turned into disadvantages, as the tables of local mortality sufficiently prove.

Extract from a local sanitary report, for the whole of which see Appendix B. (page 73).

" Hexham, although possessing every natural advantage for cleanliness and healthfulness, has, by its construction, been rendered comparatively filthy and unhealthy. The houses are much crowded in the centre of the town, so much so as to prevent the inhabitants from having ordinary conveniences (privies), or if they have them, to make their existence almost as great an evil, if not greater, than their entire absence.

" The whole of the houses that enclose the Market-place are thus situated ; on the west and north no conveniences (privies) exist ; there are no back-yards. In a few houses wooden erections are used as privies by the families, one or two having the refuse removed daily, others allowing it to accumulate until the whole neighbourhood is infected by the poisonous effluvia.

" On the south and east sides of the Market-place, behind the houses there are small yards, containing an area of 5 or 6 feet square, in which are ashpit and privy. On all sides the yards are surrounded by high walls, so that however much the wind may blow the contaminated precincts feel no breeze. Windows in many instances overlook these yards, which when opened must necessarily admit the accumulated gases that are generated beneath, with results that cannot be other than injurious.

" Close to the Market-place, extending in a southerly direction, is a range of houses dividing the two principal streets, and having a frontage into each. These houses are placed upon the smallest

---

* Since gasworks have been established, this does not apply.—R. R.

possible compass of ground, and have connected with each, one of the small yards previously mentioned. The stench from those yards at all times is represented as very offensive, but especially in summer. The conveniences (privies) are necessarily between the dwelling-houses, so that in whatever direction the wind may blow the effluvia is carried into the rooms on one side or the other.

"The ashpits, in most cases, are the receptacles of the refuse water from the houses, and consequently from the damp walls is constantly oozing a putrid semi-liquid, which slowly spreads over the yards and down the channels, increasing greatly the evaporating surface, and thereby augmenting the evil. The houses in the Fore-street are much exposed to this form of nuisance, about which the inhabitants justly complain. And, as if the sanitary condition of the locality was not thus rendered bad enough, pigs and horses are in some instances crowded into the yards.

"At the south-west corner of the Market-place debouches a passage which is the entrance to a large yard covered with the most active sources of disease. Dwellings partly surround the square, one side being occupied by that portion of the parish church which is used for religious services. There are four slaughter-houses, with their accompanying offal-heaps, always in a state of decomposition, also about a dozen piggeries, most of which have separate dung-heaps. A large stagnant drain exists, into which filthy water and vegetable and animal refuse are regularly thrown, while a large ashpit and two privies occupy the centre, where the refuse of some 15 or 20 families is daily deposited and allowed to accumulate for months. At the end of the Fore-street is a yard in many respects similar to that just described, being scarcely so large but much more filthy.

"Towards the north-west, from the Market-place, extends a range of buildings fronting Market-street; they are densely populated, and are provided with conveniences as follows:—Two privies constructed within the dwellings, emptied every morning, one in a yard barely sufficient to contain it; this is worse than the others, being so much seldomer cleaned. In a yard 12 or 14 feet square is a piggery, privy, and ashpit, which are cleaned but once or twice a year. There are two privies constructed on the public road, but kept private property. The remainder of the population, consisting of 19-20ths of the whole, have no means of getting rid of their refuse except that of carrying it to a considerable distance, which is seldom done; it is generally thrown out into the open channel or into the churchyard. All the houses are alike unprovided with means of removing refuse water. For their convenience in this respect a drain was constructed a few years ago, with an open gutter above it, but the ingenious mechanic to whom the work was entrusted formed the drain and gutter so that the water actually runs in a reverse direction to that which it ought, and in consequence they are both at all times full, and are thus rendered serious nuisances instead of public benefits; in addition, within three yards of the back doors and windows of these houses, there is a very much overcrowded churchyard, the evil influence of which it is difficult to calculate.

"Gilligate,—the same sanitary deficiencies are more apparent and even more noxious. There are several groups of houses thickly inhabited, possessing no conveniences (privies), and in a majority of instances where such exist, they are used by several families, and, consequently, no one attempts to keep them clean; the want of drainage causes stagnant pools to exist. In one instance the drainage from a piggery, privy, and ashpit has found its way into the room of an adjoining house, and the yard being level with the second floor of the house, the liquid oozes through the wall, and runs down from the second to the first floor, in such quantity as to wet the beds. A well has been sunk in the room several feet deep, in which these pestiferous drainings are allowed to accumulate until it is full, and then they are removed to make room for more. Throughout the whole of this part of the town, the want of drainage is very manifest. Its low position prevents it from having much advantage from natural drainage, so that in every back-yard exist stagnant pools.

"Black-bull-bank possesses every natural facility for drainage; five drains open on to it, and their contents flowing down, form a stagnant pool at the foot. These drains are all of an offensive character, and are much complained against by the inhabitants. The conveniences (privies) for this district are in many instances very insufficient, several families being entirely without such places. There is no means of removing the refuse water of the various families except by an open channel, which augments the evils already mentioned.

"One house is provided with an ashpit under the dwelling-room, and although it receives the contents of a privy, it is not emptied more than once a year.

"Such are a few of the more glaring sanitary evils to be found in Hexham. Many more might be mentioned. There are only about some half dozen waterclosets to be found in the town, and in consequence of the absence of drainage, they prove serious nuisances to the neighbourhoods in which they exist. The structures intended for drains are, except in very wet weather, reservoirs, so that almost constantly, from every opening, but especially from those in the vicinity of a watercloset, exhale very offensive and injurious effluvia.

"The sanitary condition of Hexham is of the lowest class. The few attempts that have been made to improve it, have been too paltry and unconnected to be of much use; they have been made for individual rather than public good, hence their inutility, and in many instances ultimate injuriousness.

"It is not necessary to prove the consequences of such a condition upon the health of the inhabitants. Long courses of experience and observation by members of the medical profession have indisputably shown that in proportion as inferior sanitary regulations exist, so is there disease in excess.

"The late epidemic of small-pox has afforded ample proof, that in overcrowded and low damp houses in the neighbourhood of offal-heaps, disease finds its most numerous victims, and there operates with the most deadly effect; many of the places referred to as being filthy, have suffered most from small-pox. In one house where there were about 40 persons residing, there have been 13 cases of small-pox, and 12 of these cases were in two families. The windows of the rooms in which these families live open upon a yard, in which are two piggeries, a filthy privy, and a large dung-heap. Over the surface of the yard flows the whole of the refuse water from all the families on the premises. There have been 10 cases in the neighbourhood of the yard at the end of Fore-street before mentioned, where slaughter-houses, piggeries, dung-heaps, stagnant putrid pools, and all the other filthy concomitants are crowded together beneath the windows of a row of houses that are parallel therewith. From evidence supplied by medical gentlemen of the town, it is conclusive that three fourths of the cases of small-pox have been in the districts pointed out in this report as being in the worst sanitary condition. And there can be no doubt physical illness is but one of the many evils consequent upon such a state of things. A clean town is an important step towards having a cleanly and moral population."

LOCAL GOVERNMENT.—Hexham is not a corporate town. Anciently the civil government was vested in the archbishop's seneschal, but afterwards in a bailiff,* who is still appointed by the lord of the manor, whose representative he is in the court over which he presides. When, in the reign of Elizabeth, the regality or liberty of Hexham and Hexhamshire were united to the county of Northumberland, the powers of its bailiff were limited. A court leet and view of frankpledge, a court baron, and two courts for the recovery of debts continue to be held periodically within the liberty. Though the town is not corporate there are four incorporated companies or trades. These are, 1st. weavers; 2d. tanners and shoemakers; 3d. skinners and glovers; and 4th. hatters.

TRADE.—There are several skinneries and tanneries in or near the town. Gloves and hats are made to a considerable extent, and there are also worsted manufactories, with a variety of other trades. The local manufactures of Hexham have, however, declined of late.

MARKETS AND FAIRS.—The principal market day is Tuesday throughout the year. There is an inferior market on

---

* I am not aware that the bailiff occupies the place of the ancient seneschal. I should rather say that Mr. Losh is the seneschal, and I am the bailiff, and that we each exercise such portions of the ancient functions of those officers as remain to be performed since the annexation of Hexham to the county of Northumberland. There is also a serjeant, who serves all summonses, and levies executions and fines, and a gaoler who has charge of the debtors prison annexed to the court of pleas.—JASPER GIBSON.

Saturdays, and a cattle market is held on alternate Tuesdays. There are fairs for horses and cattle, March, August, and November. A wool fair on the 2d of July. Hirings are held yearly ; namely, on the first Tuesday in March for hinds, and for male and female servants on the 13th May and on the 11th November.

Hotels, Public-houses, Spirit-sellers, and Beer-shops.—Hotels, public-houses, and spirit-sellers, 31 ; beer-shops, 5 ; total, 36.

Slaughter-houses.—There are 8 slaughter-houses, situated in Long-yard. It was stated at the inquiry that W. B. Beaumont, Esq., M.P. has purchased this property, to do away with the nuisance in its present position.

Baths.—There are two private baths belonging to Joseph Temple Plumber, at Bull-bank, which may be used by the public. There are no public wash-houses.

ROADS, &c.—Hexham turnpike road, under the control and management of trustees, extends through the town, and is repaired by the trustees surveyor.

The highways are repaired by surveyors chosen annually for this purpose. The present surveyors are Thomas Pratt, gentleman ; Robert Scott, gentleman ; W. A. Temperley, corn merchant ; William Robb, draper. Several highways are repaired *ratione tenuræ*. The foot-paths along the turnpike are repaired by the surveyor to the trustees ; those along the highways by the surveyor to the highways. The Newcastle and Carlisle railway passes near Hexham, where there is a station.

RENTAL OF LAND.—Land for garden purposes, near the town, lets for about 8*l.* per acre, and grass land at about 4*l.* per acre. Wages to agricultural labourers range at about 12*s.* per week.

INQUIRY.

NAMES OF GENTLEMEN PRESENT.—The gentlemen as hereunder named were present at the inquiry :—The Reverend Joseph Hudson ; William Bell, Esq. ; Messrs. Jasper Gibson, solicitor ; Richard Gibson, solicitor ; Charles Head, solicitor ; Henry Dodd, solicitor ; John Stokoe, solicitor ; Thomas W. Welford, solicitor ; John Taylor, solicitor ; Robert Pattinson, solicitor ; S. Pemberton, solicitor ; William Kirsopp, solicitor ; Thomas Jefferson, surgeon ; Thomas Stainthorpe, surgeon ; Robert Stokoe, surgeon ; William Pearson, surgeon ; John Charlton ; Joseph Fairless ; Smith Stobart ;

Inquiry.

James Turnbull; William Taylor; Wylam Walker, civil engineer; John Ridley, farmer; George Robinson, draper; Matthew Ord, silversmith; Matthew Smith, farmer; Joseph Ridley, glover; William Wilson Gibson, chemist; William Pruddah, chemist; William Robb, draper; Henry Walton, actuary; William Ellis, painter; Thomas Clemitson, merchant; Thomas Dinning, draper; and others.

PRELIMINARY REMARKS.—As in other places which I have visited, I found much excitement in the town of Hexham relative to the inquiry. This was evinced immediately upon my attempting to take evidence. The room was tolerably crowded, and the promoters, as also the opponents were present. At the inquiry, however, and during my subsequent inspection of the district, I received the utmost assistance from all parties, and my thanks are especially due to the Messrs. Gibson, Charles Head, John Stokoe, and others. The documentary, historical, and other evidence furnished is embodied in this Report and in the Appendices.

Mr. Ralph Errington Ridley having proved the publication of the notices as the Act directs, Mr. John Stokoe, solicitor, protested against the inquiry, and wished to examine the petition. This I allowed him to do, and after some irrelevant conversation, it was agreed that the inquiry should proceed, the petition having been found genuine and legal. Mr. Charles Head opened the case of the promoters, and denied that there had been any "hole-and-corner meetings," but public meetings were called, at which the question of applying for the Public Health Act was discussed. There had been two fires some twelve months previously, which showed the want of water; and public hand-bills had been posted calling a meeting to consider the propriety of seeking legal power to extend the waterworks, but the heavy cost of a private Act of Parliament deterred the promoters of the meeting from taking such a course.

The answers to the inquiries under the several heads were as follows:—

There are no *Local Acts of Parliament* in force for paving, lighting, cleansing, watching, regulating, or for supplying with water, or for improving the town of Hexham.

Boundaries.—Hexham township is bounded on the north by the river Tyne; on the east by the parish of Corbridge: on the south by the parish of Corbridge, the township of Hexhamshire low quarter, and the township of the west quarter of Hexhamshire; and on the west by the township of the west quarter of Hexhamshire and the parish of Warden.

Inquiry.

Area.—The area of Hexham township is abont 4,596 acres. 4,413 acres pay to the county rate. The area upon which the town stands is not known.

Population of Hexham Township.—

1801	3,427
1811	3,512
1821	4,116
1831	4,666
1841	4,742
1851	5,231

In 1821, with a population of 4,116, there were 1,801 males, 2,315 females, 529 houses, and 1,028 families. In 1841, the population consisted of 2,168 males, and 2,574 females. There were 690 inhabited houses, 29 uninhabited houses, and 5 houses in course of erection. In 1851, there were 2,247 males, 2,804 females; 1,191 separate occupations; 667 inhabited houses; 22 uninhabited houses, and 1 building.

Estimated annual value of land and buildings, as estimated in 1809 for the county rate to new gaol, 8,350*l.*

Sewerage and Drainage.—There is no proper system of sewers and drains (see report in Appendix, p. 73); such sewers as have been made are of rubble, square in section, and are not adapted for house drainage. The amount of money expended in making these defective sewers could not be furnished; they had been made from time to time, and had been paid for out of the highway rates.

Water Supply.—The present water supply is described more fully at pages 66 to 71. Mr. *Matthew Smith* stated that he owned four wells or springs of water, which supply the town. Mr. *Joseph Fairless* stated that there was one fire-engine, but neither this nor the supply of water was adequate to the requirements of the district.

Lighting.—The town is lighted by 55 public gas-lamps. (See particulars of gasworks, p. 30.)

State of the Burial-grounds.—It was stated that no adequate means for burial existed (see statements and report, p. 77). The new burial-ground at the Roman Catholic chapel was the least crowded. In 1841 it was stated great efforts were made to provide a cemetery, but without avail. It was agreed that a new cemetery was much needed.

Number of Inhabitants.—The number of inhabitants in 1851 was 5,231 (2,427 males, 2,804 females).

Local Rates. 21

Annual Rateable Value of Property in the whole township, 11,130*l*.

Poor's Rate for year ending 29th September 1852, was 3*s*. 5*d*. in the pound; four rates, 9*d*., 10*d*., 11*d*., 11*d*. The sum expended in poor relief for the nine years ending March 1852 amounts to 17,370*l*. 18*s*. 11¼*d*.*

Highway Rate.—Year ending March 1852, a rate of 4*d*. in the pound raised 156*l*., levied upon the town portion, or "ancient lands." The "fell" portion of the township is separately rated.

Lamp and Watch Rate.—Year ending 29th October 1852, rate levied on the whole township; houses, 7½*d*. in the pound; land, 2½*d*. in the pound; amount raised, 248*l*.

Church Rate.—Church rate is a separate rate, but at present it is collected by the collector of the poor's rate. The assessment is 1½*d*. in the pound; producing 66*l*. 6*s*.

Sanitary Condition of the Inhabitants.—The sanitary condition of the inhabitants is such that the annual mortality averages, according to Mr. John Stokoe, upwards of 25 in 1,000, and according to another return 29¼ in 1,000. Cholera visited the town in 1832, and again in 1849; fever prevails.

Rateable Value of Hexham Town.—

		Numbers.
5*l*. and under	-	306
Not exceeding 10*l*.	-	300
,, 15*l*.	-	164
,, 20*l*.	-	98
,, 30*l*.	-	54
,, 40*l*.	-	26
,, 50*l*.	-	10
,, 60*l*.	-	1
,, 70*l*.	-	4
,, 80*l*.	-	2
,, 90*l*.	-	3
,, 100*l*.	-	1
,, 150*l*.	-	1
,, 200*l*.	-	1
,, 500*l*	-	1

	£	s.	d.
Annual rateable value, according to rate book, in June 1852 (last rate), for Hexham poor rate and other rates	11,008	7	6†
,, ,, county rate	12,515	0	0

---

* Those who oppose improvement, in dread of the expense, should consider the excessive poor's rate. Proper works of sewerage and of water supply would reduce this charge, diminish sickness, and prevent premature deaths.

† When accounts audited by auditor, 11,005*l*. 3*s*. 6*d*.

		£	s.	d.	
Amount of church rate for last year		66	7	5¼	for year.
Land tax	- - -	54	5	7	,,
Land tax redeemed	- -	52	9	5¾	,,
Lamp and watch rate	- -	248	14	6½	,,
House tax and assessed taxes	-	430	10	3	,,
Property and income tax	- -	679	7	1	,,
Poor·rate	- - -	1,880	7	4	,,
Highway rate	- -	156	13	2	,,
Yearly rates, cesses, and taxes	£	3,568	14	10½	

	£	s.	d.	
Tithe rentcharge	550	0	0	
Lords rents	- 21	18	7	for two manors in the town, Hexham and Anick Grainge.

EVIDENCE OF MEDICAL GENTLEMEN AND OTHERS.—Mr. *Thomas Jefferson,* surgeon, stated,—

"Knows Hexham well; saw cholera in 1831 and 1832, only saw one case. There is generally more or less fever, but this is not peculiar to Hexham, nor does it prevail more in one part of the town than another. Thinks if the Public Health Act was in force the sanitary condition of the town could not be improved."

Mr. *Robert Stokoe,* surgeon, stated,—

" Have been in practice 40 years ; have attended at the dispensary upwards of 30 years, of which I am now the senior surgeon, and had charge of the poor several years. Epidemics are common, endemics are not common in the district. There is generally more or less of small-pox and scarlet fever. I have always considered the town healthy, being seldom visited with contagious or infectious diseases, and never of long duration. A strong proof of the town being considered healthy, is the number of families resorting to it from Newcastle, Shields, Sunderland, and various other places ; many of the people being sent by their medical advisers. Local deaths are increased by the influx of sick strangers, four old invalids having died in one lodging-house within the last three years. I consider the list of deaths laid before you to-day to be considerably above our general average, and that this has been caused by certain epidemic diseases prevailing generally during the period when the list was taken. The epidemics prevailing were scarlet fever, which was very general in the town and neighbourhood ; the worst cases being out of the town ; small-pox prevailed very generally, chiefly amongst the non-vaccinated."[*]

Mr. *William Angus Temperley,* corn merchant,—

" Believes that Hexham is situated in a very healthy district ; but considers that Drs. Jefferson and Stokoe have attempted to

---

[*] It has been said that no physician above 40 years of age would believe Harvey's theory as to the circulation of the blood. The senior surgeons in Hexham will not believe in sanitary improvement.

prove too much; as, if the town is so healthy as they describe, the average mortality ought to be lower."

Mr. *William Pearson*, surgeon,—
" Has been 12 years in practice; has been surgeon to the union three years; finds most disease in the distressed parts of the town amongst the poor; has no doubt, but that if the filth were removed, and pure air was allowed to circulate, there would be less sickness, and less parish relief required. There is a fever ward at the workhouse, and the guardians have provided a proper conveyance for fever patients."

Mr. *Thomas Stainthorpe*, surgeon, and medical officer to the Union,—
" Considers the town very unhealthy. Attended 217 cases of small-pox during the last winter. The guardians removed a wife and four children from Victoria-place, and had the bedding burned."

Mr. *John Nicholson* and Mr. *J. B. Maughan*, surgeons, confirmed the evidence of Mr. Stainthorpe as to the defective sanitary state of the town, and the amount of "*preventible disease*" consequent thereon.

Mr. *Stainthorpe* put in the following statement:—

" *Local Mortality.*

" Sir,—According to your request, I beg to hand in the following extracts which I have taken from my medical weekly reports to the Guardians of the Hexham union, extending over a period of two years and a half. I may remark, that these reports were made week by week for the guidance of the Board of Guardians in administering relief to the sick and the families depending upon them, and not for the purpose of making out a case or yet colouring the picture too darkly; for I am confident that I am within the truth. It is also essential that I should remark, that the accompanying Tables do not contain a single case of sickness of the inmates of the workhouse, excepting those cases of typhus fever occurring in the town, and which I have seen it would be necessary to remove to the fever wards of the workhouse. They necessarily appear first in the columns of my district medical report book, and afterwards in the workhouse medical report book.

Inflammatory, remittent, continued, and typhus fevers	75 Cases.
Scarlet fever	22
Small-pox	56
Diarrhœa (including English cholera)	38
Dysentery	5
Influenza	24
Erysipelas	4
Measles	7
Hooping-cough	1
Asiatic cholera	1
	233

24 Localities of epidemic and contagious Disease.

"The following Table was drawn out immediately after the epidemic of small-pox which prevailed in Hexham during the last winter, 1851 and 1852, extending over a period of about four months, and sent to each medical gentlemen named in the columns, so that the figures were inserted by themselves. I have much pleasure in stating that the medical gentlemen at once accommodated me, and with much cordiality.

CASES of SMALL-POX last Winter, during four months.

Names of places where epidemic, endemic, and contagious diseases prevail.	W. Stokoe.	R. Stokoe.	T. Jefferson.	W. Pearson.	J. Nicholson, and J. B. Maughan.	J. Farbridge.	Thomas Stainthorpe.	Total
Gilligate	–	5	1	1	7	2	49	65
Barracks (in Gilligate)	–	–	–	–	–	–	2	2
Dean-street	–	–	–	–	–	–	2	2
Fore-street	–	–	–	1	1	–	–	2
Cockshaw	–	2	1	–	–	2	5	10
Hall-gate	–	–	–	2	1	–	4	7
Cattle-market	–	–	–	1	2	3	–	6
Old Burn-lane	–	–	–	–	–	–	2	2
Churchway	–	–	–	–	–	–	4	4
Tyne-green	–	–	–	2	2	–	1	5
Bull-bank	–	–	–	1	2	–	–	3
Skinners-burn	–	–	–	–	4	2	15	21
Priest-popple	–	1	–	1	1	–	–	3
Back street	–	1	–	3	1	5	3	13
Broad-gates	–	–	–	–	3	6	–	9
Battle-hill	1	–	1	–	–	–	6	8
Market-place	–	1	1	–	3	–	–	5
Victoria-place (Battle-hill)	–	–	–	–	1	–	5	6
Helensgate-lane	–	–	–	–	–	–	4	4
Haugh-lane	–	2	–	–	4	–	1	7
Workhouse (Hexham Union).	–	–	–	–	–	–	20	20
Hencotes	–	3	–	1	–	–	–	4
Market-street	–	1	–	–	–	3	–	4
House of Correction (near Tyne-green).	–	1	–	–	–	–	–	1
West Spital lodge (a mile from Hexham).	–	3	–	–	–	–	–	3
	1	20	4	13	32	23	123	216

SMALL-POX CASES, arranged into WARDS.

Gilligate Ward, including Gilligate, Cockshaw, Old Burn-lane, Tyne-green, Helensgate-lane, Haugh-lane, and House of Correction - - 96

Priest-popple Ward, including Priest-popple, Dean-street, Cattle-market, Broad-gates, and Skinners-burn - - - - 41

Market-street Ward, including Market-street, Fore-street, Hall-gate, Church-way, Bull-bank, and Market-place - - - - **25**

Hencotes Ward, including Hencotes, Back-street,
Battle hill, Victoria-place, and West Spital cottage - - - - - 34
Hexham Union Workhouse, (many of these cases were removed to the fever wards of the workhouse from the different localities of the town) - 20

216

TABLE OF FEVER CASES.

"The following Table of 71 cases of fever, during the last 2½ years, shows the locality most affected with endemic, epidemic, and contagious diseases:—

Gilligate	- 34		Broad-gates	- 2
Cockshaw	- 2		Bow-bridge	- 3
Haugh-lane	- 1		Back-street	- 3
Helensgate-lane	- 2		Battle-hill	- 3
Old Burn-lane	- 3		Hencotes	- 3
Tyne-green	- 2		Long-yard	- 2
Fore-street	- 2			—
Skinners-burn	- 5			71
Church-row	- 2			—
Priest-popple	- 2			

"I have been medical officer of the workhouse and district of Hexham for 9½ years, and believe the foregoing to be something approximating, as near as possible, the regular state of my experience during that period, with the exception of small-pox, which prevailed so alarmingly during last winter. I should suppose that not more than two cases yearly occurred during my former experience. In February 1849, five or six cases of Asiatic cholera occurred in Hexham, and I believe the town was saved at that time by a general meeting of the inhabitants held in the Moot-hall, forming a committee of medical gentlemen to make domiciliary visits for the purpose of administering medical relief, general relief in the shape of food, bed and body clothing, and causing the town to be thoroughly cleansed, by giving *hot lime*, and lending brushes and other necessaries for the required purification.

(Signed) " THOS. STAINTHORPE."

" Battle hill, Friday morning.

" N.B —It occurred to me this morning, that the following information might be useful:—

" In 9½ years the mortality in Hexham union workhouse, ending September 1852, was 207 cases ; 69 of the above 207 belonged to Hexham ; 26 cases belonged to the establishment, or common charges account (chargeable to the entire union); Hexham pays about one fifth to that fund, so that in fairness one fifth of the mortality ought to be added to Hexham.

" THOMAS STAINTHORPE."

MORTALITY IN HEXHAM TOWNSHIP.

	1841.	1851.
Population in Township	4,742	5,231
Population in Town	4,161	4,601

Years.	Deaths.	Average per 1,000.
1845	106	21·20
1846	149	29·17
1847	156	30·73
1848	116	22·67
1849	145	28·13
1850	124	23·88
1851	114	21·79
		7)177·57
Average of seven years	-	25·367

N.B. The pauper deaths from other parts of the parish are not included :—

1841, the town population } in prison - 7
includes - - } in workhouse - 121

Total - 138

1851, the town population } in prison —
includes - - } in workhouse 185

Total - 185

DEATHS in HEXHAM TOWNSHIP, exclusive of WORKHOUSE, up to 30th June 1852.

Hexham township, including those in workhouse
charged to Hexham - - - - 933
Chargeable to other townships - - - 80

1,013
Total deaths in workhouse - - - 154

7) 859

122⅚

AVERAGE DEATHS of HEXHAM TOWNSHIP, exclusive of the WORK-HOUSE, up to 29th September 1852, for seven years.
Population - - - 5,231.
Average deaths - - 23·70 per 1,000.

Statistics—Poor Relief.

POOR RELIEF.—Hexham union comprises 70 parishes and townships, including Hexham township; and elects 82 guardians, exclusive of the magistrates, who are ex-officio guardians; four relieving officers. The total population in the various parishes in the union, according to the last census in 1851, was 30,436.

	£	s.	d.
Total amount of the expendiditure of the union for the half year ending Michaelmas 1851	4,683	18	10¾
Half year ending Lady-day 1852	4,721	6	9¾
Exclusive of overseers accounts and expenditure	9,405	5	8½

Of this amount the township of Hexham expended:—

	£	s.	d.
Half-year to Michaelmas 1851	849	2	3¼
„ Lady-day 1852	938	17	4¼
	£1,787	19	7½
Amount expended by overseers for Hexham for the year ending Lady-day 1852	118	1	2
Total expended by Hexham township	£1,906	0	9½
Amount expended by overseers for the different parishes and townships in union	458	6	9¼
Brought down	9,405	5	8½
Total expenditure in union for last year, ending Lady-day 1852	£9,863	12	5¾

INSPECTION OF TOWN AND DISTRICT, *Sept.* 30, 1852.

Names of gentlemen who attended during the inspection:—Messrs. Jasper Gibson, Richard Gisbon, John Charlton, Wylam Walker, Charles Head, Ralph Errington Ridley, Thomas Jefferson, Thomas Pratt, William Wilson Gibson, William Robb, and others.

Abstract of notes taken on a personal inspection of the town and district, with remarks.

Water Supply.—The present supply of water is partly public and partly private. A pipe of three inches diameter brings water from a brook or spring called Hencotesburn, where rude and very imperfect works exist. The delivery is said to be about 24,000 gallons during each

24 hours. Within the town the water is delivered into "pants," open troughs, and lead cisterns, from which it is taken by the inhabitants; much of this limited quantity, consequently, runs to waste. The upper reservoir is about 70 feet above the pant in the Market-place. There is no filtering apparatus in the brook, and consequently, during wet weather, the water becomes turbid, and it is delivered in this state in the town. The morning of my inspection was wet, and the water was in a very filthy state. I examined the source of the present supply, and the district around, and found several natural springs, bright, and comparatively pure at all times. Spring-water in abundance may be collected at such an elevation as will supply the whole town. Mr. William Pruddah stated that in 1826 (an exceedingly dry summer) he had seen these springs, and they were plenteous during the whole year. The present water supply is limited, and occasionally very foul. A supply may be obtained abundant and pure. Certain tanners and skinners use the water of the brook, but an impounding reservoir may be made for their compensation. A considerable addition may be made to the spring water by deep drains.

Drainage, &c.—On inspection I found the sewerage and drainage of Hexham most defective. The houses are generally confined; and, in many cases, the slope of the land is towards the streets. There are privies with open cesspools and large open middens behind almost every house, or group of houses. This is so common, that I almost agree with Mr. Jefferson, that there are not in Hexham any places or district where epidemic diseases especially prevail. The whole town must be liable to such diseases.

Condition of Houses, &c.—Previous to this inspection some of the large middens had been emptied, and many of the yards and streets had been cleansed, so as to remove, as much as possible, evidences of the general filth; but the faulty structural arrangements remained, and proper sewers and drains could not be as suddenly swept in. I have found this spasmodic mode of cleansing to deceive the inspector adopted in other places I have visited.

House, No. , Temperley-place.—Water was standing one foot in depth in the cellar of this house; it has to be baled or pumped out.

Remarks.—Many families live in single rooms, the rent of which varies from 6*d.* to 1*s.* per week. There are few single cottages within the town. On the outside of the town cottages having two rooms let from 4*l*. to 4*l*. 10*s*. a year. Those at 5*l*. generally have a small garden attached.

Old Golden-Lion-yard.—There are several room tenements adjoining this yard. There is one privy, the seat is however generally so dirty as to be unfit for use. There is a large open cesspool attached, the landlord claiming the manure and removing it as may suit his convenience. The tenants throw their slops out on to the surface of the yard; several of them complained of the inconvenience and general filth.

Victoria-place.—There are common lodging-houses in this place. The common stairs to the several tenements are of wood, with lath and plaster partitions; so that should a fire occur, the whole must be burned down, to the danger of many lives. There is one privy and large midden behind.

White Swan, or Gaping-goose-yard.—The yard is unpaved, and is rendered filthy by pigsties, foul middens, and adjoining slaughter-houses.

Slaughter-houses.—There are several slaughter-houses near the church; they are generally very filthy. The yards are confined. Persons in church are said to be frequently annoyed by the foul smell from these places.

Girls Sewing-school.—The room is low. The yard is unpaved. There is a privy and large open midden near. The school is usually crowded with children.

Battle-hill.—Opposite registrar's office. The yard is confined, and contains a privy and midden, usually foul and offensive.

The streets in Hexham are part Macadam-stone, part pebble paving, and part square sets. The footwalks are partially flagged and partly paved with pebbles. The houses are of stone and of brick; some are rough-cast. The streets are irregular in line, and the houses are very unequal in style and in elevation. Old houses are thatched; others are covered with the heavy grey slates of the district; most of the new ones are covered with blue slates.

The Seal.—This is an open area of grass land of about 20 acres, and is used as a recreation ground. It belongs to W. B. Beaumont, Esq., M.P. This large open area must be of singular advantage to the public of Hexham.

On the lower part of Hexham there is a considerable breadth of land from Haugh-lane to the river Tyne, used as nursery ground, and for market gardens. To this land much of the town's refuse might be applied.

I was informed that nursery-men and farmers have intimated their readiness to lease the town's refuse if it were all collected to some available point.

Lighting and Watching.—The Act 4 Geo. 4. was adopted for watching the township of Hexham, and lighting it with oil lamps, 9th December 1830.

Act 3 & 4 Will. 4. cap. 90. was adopted for lighting only, 14th October 1833; adopted for both lighting and watching, 29th October 1834; which up to this time is in operation.

The first contract for lighting the town with 46 gas lamps, dates 21st September 1835.

Nine inspectors are appointed, three of which in rotation go out of office annually.

At the present time (September 1852) there are 55 street lamps in Hexham, which, with posts, lamps, and everything belonging thereto, are the property of the Gas Company; and they contract with the inspectors to provide a lamp-lighter, and light these lamps at the rate of two guineas per year for each lamp, burning 1,250 hours.

The lamps are generally lighted from the latter end of August or beginning of September to the April or May following, according to the moon; that is, they are lighted two or three days after every full moon, and continue until the next moon is one quarter old; sometimes more, according to the weather.

(For Statements of Accounts, see pp. 31, 32.)

" N.B.—In making a report respecting the gasworks at Hexham, I may state to you that a company was formed, consisting of 270 shares of 10*l.* each.

" The works were established in 1834, and the number of retorts at that day consisted only of 4; and in 1838 there were 9 retorts. In 1850 the retort house was enlarged, and then there were ovens made to hold 13; and they will consequently make more gas at any time than we consume. Our purifier is at present too little for what we make, but ultimately will have to be made larger. Our gasometer is the same as we put in at the first, and we are in the same position in respect to it. Our gasometer and other apparatus are beginning to be too little for the quantity consumed. We have 55 public lamps, for which the inspector of the town pays the Commissioners at the rate of two guineas per lamp, on condition that they burn 1,270 hours, their secretary giving us a scale of the burning for each month. I may likewise state, that there is no lighting in summer, when the days are at the longest,—say from 11 to 13 weeks.

" Price at first, and up to 1846, 10*s.* per 1,000 feet; present price, 8*s.* 6*d.* per 1,000 feet. Main-pipe 4 inches diameter."

STATEMENT of the Receipt and Expenditure of the Lamp and Watch Inspectors, for seven years.

	RECEIPTS.					PAYMENTS.				
—	From rate.	From fees received by officers for serving summonses and for fines.	From persons for use of fire-engine out of the township, &c.	Other receipts.	Total.	For lighting lamps.	For police.	For fire-engine.	Secretary and treasurer's salaries, stationery, &c.	Total.
	£ s. d.	£ s. d.	£ s. d.	£ s. d.	£ s. d.	£ s. d.	£ s. d.	£ s. d	£ s. d.	£ s. d.
Balance in hand, 29th Oct. 1844	- - -	- - -	- - -	- -	11 10 5¾					
Years ending 29th Oct. 1845	221 10 0	- - -	2 2 0	- -	223 12 0	96 6 8	99 8 11½	6 19 3 ‡15 11 3	§14 4 3	232 9 4½
1846	231 4 0	- - -	- - -	- -	231 4 0	96 0 0	101 5 4	14 16 2	§17 12 8	229 14 2
1847	233 9 4¼	12 12 5	- - -	- -	246 1 9¼	94 10 0	108 4 3	14 3 10	§13 14 10	220 12 11
1848	225 10 5½	7 18 5	2 10 0	†0 19 6	236 18 4½	106 16 0	112 2 0	17 10 5	9 12 4	240 0 9
1849	207 11 0	10 4 3	6 7 0	- -	224 2 9	106 1 0	107 18 7	6 15 5	7 14 6	228 9 6
1850	234 12 0	10 12 4	1 7 0	- -	246 11 4	109 4 0	108 6 6	6 19 5	7 10 11	232 0 10
1851	230 5 8	4 5 1	7 8 0	- -	241 18 9	110 9 0	117 6 0	16 12 7	8 10 7½	252 18 2½
Balance in treasurer's hands, 29th Oct. 1851	- - -	- - -	- - -	- -	- - -	- - -	- - -	- - -	- - -	25 13 8½
	*1,584 2 5¾			£	1,661 19 5½				£	1,661 19 5½

* 7) 1,584 2 5¾

£226 6 1 Average annual amount raised by rate.

† Old engine-pipes.
‡ Building engine-house.
§ These amounts include rent of lock-up house and office for inspectors meeting.

GENERAL STATEMENT of the Income and Expenditure of the Hexham Subscription Gaslight Company, between the 19th day of December 1849 and the 18th day of December 1850.

INCOME.		£	s.	d.
To balance in company's favour		136	6	4
Arrears of gas rent		10	19	5
Received for coke		2	1	8
" tar, lime, &c.		34	9	2
" coke and tar sold at yard		8	7	7
		1	5	11
By Contract.				
First quarter's gas rent, due Jan.	48 4 2			
Second ditto April	14 19 8			
Third ditto July	14 5 4			
Fourth ditto Oct.	38 10 5			
		115	19	7
By Meter.				
October 1849 to January 1850	111 14 0			
January to April	88 0 8			
April to October	46 14 8			
		246	9	4
Received from lamp inspectors		109	4	0
Rent of meters		0	6	0
Rent of garden		3	10	0
		£668	19	0

EXPENDITURE.		£	s.	d.
Amount paid on dividends		108	0	0
Paid for coals	99 11 9			
Leading ditto	11 19 8			
		111	11	5
Superintendent's salary	20 0 0			
Stoker's wages	43 2 0			
Lamplighters	14 16 0			
		77	18	0
Enlarging Gas-house and resetting Retorts.				
Masonry	41 15 7			
Fire-bricks	37 5 0			
Abbot and Co. for 8 retorts, &c.	38 3 11			
Blacksmiths	24 18 1			
Plumbers, &c.	5 7 9			
Leading	13 12 0			
Slating	8 2 6			
Sundry small accounts	5 16 2¼			
		175	1	6¼
Masonry account unpaid		41	15	7
Leading and labourage		133	5	5¼
Rates, income tax, &c.		29	7	7
Blacksmith, plumbing, &c.		21	12	1
Tradesmen's bills		22	17	11
Incidentals		35	14	0¼
Balance in favour of the company		3	13	10¼
		124	18	7¼
		£668	19	0

3d *January* 1851. Examined and found correct.

EDWARD PRUDDAH, } *Auditors.*
ISAAC BATY.

REMARKS.—The price charged for gas in Hexham is high, considering that coal is cheap. An excessive charge is alike injurious to the maker as to the consumer, and sooner or later begets discontent and rivalry. A moderate charge induces general consumption without very much increasing the establishment charges. Gas may be made and sold to a profit in Hexham at a price not exceeding 4s. or 5s. per 1,000 feet. In Whitehaven the price is 4s. per 1,000 feet,* in many other places it is as low, in some even lower. If the company study their interest they will extend their works, enlarge their mains, and reduce the price.

BURIAL GROUNDS.—The following report from Mr. Fairless proves that a new burial-ground is required. A plan of the yard is given in this Report, which shows the crowded state of the district; there are slaughter-houses, pigsties, privies, and dwelling-houses.

" A REPORT of the AREA and STATE of the BURIAL GROUND at HEXHAM, 1852.

" The church and burial-ground lie nearly in the centre of the town, although not in close proximity to the houses, except on its north-east side, which is bounded by the Market-place and one side of Market-street. It is partially open to the west, with the exception of the modern buildings standing on the site of the ancient abbey and grounds, which are rather spacious. The burial-ground must have been of greater extent in early times, as we find the foundation of the adjacent houses strewed with human bones. We have no trace in history of the extent and population of this ancient town, but what is connected with its early church, which rose into existence in the seventh century, along with its *cathedral*; there were besides *two parish churches*, traces of which still remain; hence we may infer a greater population and extended space of sepulture, such as I have just stated.

" The area of the present occupied churchyard is in measurement 3 roods and 2 perches, and the soil is remarkably dry at the depth of interment.

" The average of burials for seven years ending 1851, gives the number of 127 annually; and allowing the space of two square yards to one grave, would admit of a period of fifteen years before the necessity would arise of breaking into a former grave.

" *Burial within the Church.*—Right of interment within the walls of the church is claimed by a great number of families in the town; as many as 40 places can be pointed to where different families have buried. The power of granting and investigating the right of claim was held by the ancient ' twenty-four' or select vestrymen, but it has dwindled away and fallen into the hands or will of the incumbent.

" The last seven years give over 20 interments in the church.

---

* Gas, in Whitehaven, has been reduced to 2s. per 1,000 feet since the inquiry, according to published statements.

"The crowded state of the churchyard is generally admitted, and interments within the church highly reprobated, consequently some change is very desirable.

(Signed) "Jos. Fairless."

Total space of burying-ground, 3,892 square yards.

Burials in the Ground attached to the Roman Catholic Church of St. Mary's.—The following return, obtained by Mr. Gibson, was handed in at the inquiry:—

"St. Mary's Hexham,
"My dear Sir, "28th September 1852.

"According to your request, the following are the numbers that have been buried in St. Mary's Roman Catholic Cemetery, since the year 1844, viz.:—

"In 1845, 7; 1846, 9; 1847, 8; 1848, 5; 1849, 5; 1850, 2; 1851, 6.—Total, 42.

"The above I have faithfully copied from the register kept by me. "Michael Singleton."

"To J. Gibson, Esq., Hexham."

Fees for Marriages, Burials, &c. in the Parish Church.— The following table of fees was furnished by the Rev. J. Hudson, incumbent:—

"Sir, "Hexham, 30th September 1852.

"In compliance with your request, I have much pleasure in furnishing you with a table of church fees paid in the parish of Hexham.

		s.	d.
For every licensed marriage - to the minister -	10	0
,, ,, - to the clerk -	4	0
For every marriage by banns - to the minister -	1	3
,, ,, - to the clerk -	0	6
For every publication of banns - to the minister -	1	0
,, ,, - to the clerk -	0	6
For every funeral in the churchyard to the minister -	0	4
,, ,, - to the clerk -	0	3
For every funeral in the church or ⎰ to the minister -	0	8
from another parish - ⎱ to the clerk -	0	6
For every churching of a woman - to the minister -	0	6
,, ,, - to the clerk -	0	2

"I may add that it is the custom for the clerk to charge for tolling the bell at each death 1s., and at each burial 6d.

"No charge has been made by me for the erection of tombstones in the churchyard since I entered upon the incumbency in 1845, as I understood that none had been made by my immediate predecessor, though I have been told that something was formerly paid for the permission. Two or three mural tablets have been put up in the church during that period; but I have made no demand, except in one case in which the party did not belong to the parish.

"The fee for each certificate of a register is 1s.

(Signed) "J. Hudson, Incumbent."

Abstract of Report.

ABSTRACT OF THE REPORT, AND SPECIFICATION OF WORKS REQUIRED.—Hexham is a small market town, situate on the south bank of the river Tyne, the substrata being millstone grit. The site of the town, as also the land generally in the neighbourhood, slopes towards the river. Two small watercourses, the *Hextol* (or Cockshaw-burn), and the *Halgut* (or Cowgarth-burn), enter the upper part of the town, and after passing the *Seal*, become *Seal-burn* down to the Tyne. The town is irregularly built. *Priest-popple*,* *Hencotes*,† and *Battle-hill*,‡ form one street, having a gradual ascent from east to west, to the south of the town immediately under the *fell*, or rising ground. The houses in this street are mostly of modern date. Bond-gate, under which flows Skinner-burn or East-burn, Broad-gates, Fore-street or Coastley-row, Back-street or St. Mary's Chair, the Marketplace, Gilligate or St. Giles's-street, the suburb of Cockshaw, Pudding-row, Hall-stile, &c., form the old part of the town. The suburb of Cockshaw is intersected by the burn, and its banks are crowded by old houses, beneath some of which the water passes. There are several narrow streets in this district, the brook being crossed by small bridges. It is in this district that most of the tanneries and glove manufactories are situated. There has not been any rapid increase in Hexham, considering its proximity to the Newcastle coal-field, and its many natural advantages. In 1743-4, Mr. Hutchinson estimated the number of inhabitants at 2,000; in 1801, the numbers were found to be 3,427; and in 1851, 5,231. Many places in England, not so favourably situated, have increased more than twentyfold within the period here named. The pure spring water to be found so abundantly in the millstone grit, above Hexham, might be used with the greatest advantage for many manufacturing purposes, as this water will be found especially valuable and economical for brewing or for dyeing.

It is shown in this Report that the mortality in Hexham is very high for such a district, (25·90 per 1,000), being above the average of all England, (22·30 per 1,000); and it is also shown that the causes leading to this excessive

---

* *Priest-popple;* the ancient district set apart by the abbot, or given as a residence to the officers and servants of the church.

† *Hencotes;* the district or place where the poultry belonging to the priory were kept.

‡ *Battle-hill;* supposed by some to be the site of a former battle. Mr. J. Gibson suggests that the name may be a corruption of the word "*Bottle*," a building, and that the name "Bottle-hill" means a hill with buildings, as Bottle-bank in Newcastle, in Gateshead, and in other places.

mortality are not natural, but structural. Many of the streets are narrow, and ill paved; most of the houses are old, and confined at the back; and, being let off in separate rooms, as tenements, they are unduly crowded; human and other refuse is stored for long periods in close proximity to the dwellings, so that the town atmosphere is poisoned with that which would be most valuable if applied to the agricultural land of the district. There is no local governing power capable of grappling with the evils which have accumulated with the growth of the place; and it is a lamentable fact that the attempt to obtain this power by the cheapest accessible means is opposed. I have examined into the sanitary state of many towns in England, from Plymouth to Berwick-upon-Tweed, from Sunderland to Liverpool, and in no town of a parallel population have I found more filth, overcrowding, and general neglect, than in Hexham. The plan accompanying this Report exhibits the general arrangements of the *burns*, streets, courts, yards, houses, privies, middens, pigsties, slaughter-houses, and grave-yard. It is ever painful to witness the concurrence of so many conditions favourable to the generation of disease, but it is more painful to find this state of things defended; and in Hexham this is the case; even medical men, who are usually the most active promoters of sanitary reform, have, in Hexham, come prominently forward, and, by their evidence, attempt to perpetuate this state of things. In this respect Hexham is exceptional; throughout my labours generally, I have ever received the most valuable countenance and aid from local medical gentlemen. As I believe that this opposition can arise only from misconception of the Public Health Act, and the cost of sewerage and of a water supply, it may be useful to explain some of the powers and provisions of the Act, as also the income and local benefit which may be derived from proper works.

The Public Health Act, 1848, provides, at the least cost to the ratepayers, for the local election of a representative body, having legal powers to efficiently sewer the town and drain the houses; to form and pave the streets; to provide a proper supply of water for public and for domestic use; to light, watch, cleanse, and otherwise regulate the district. The Legislature have ordained that this, the Public Health Act, may be obtained without incurring all the legal expenses of a private Bill, the average cost for which generally exceeds 2,000*l*. in the Commons alone. The average cost to apply the Public Health Act to such places as Hexham having been below 100*l*.

Specification of Works required.

SPECIFICATION OF WORKS REQUIRED.—*Sewerage and Drainage.*—The town of Hexham requires to be sewered and drained, but if proper means be used, and the natural advantages of the site are made available to the fullest extent, this need not be expensive, but the contrary. Efficient sewers and drains, economically carried out, add value to the property of the district, by as much as they add to the health and available comforts of the inhabitants. In Hexham the refuse of the town may be made to pay the greater portion, if not the whole, of the sewers rate. In Rugby, with a population of about 7,000 inhabitants, the Local Board have leased the refuse for 20 years, at a premium of 200*l*., and 50*l*. rent per annum, the tenant providing 1½ acres of land, manure tanks, and a ten-horse steam engine. This property reverting to the Local Board at the termination of the lease.

Waterworks.—Waterworks may be established under the powers of the Public Health Act, which shall, at the prices charged by companies, pay a 10 per cent. dividend upon the capital expended. As, however, the works would be the property of the ratepayers, either the rentcharge for water might be reduced or the surplus revenue could be laid out in town improvements.

With proper sewers and drains, a full supply of pure water, well paved and regularly cleansed streets, courts, and yards, Hexham may be made one of the most healthy towns in the north of England, as its site is one of the most beautiful. Judging from known facts, the local rate of mortality need not exceed 10 or 12 per 1,000; at present, as is shown by the returns, it is upwards of 25 per 1,000. In the Island of Portland there is a colony of convicts, about 800 in number. They are regularly worked, have few luxuries beyond fresh air without and within doors, and the average annual mortality is only at the rate of 4·5 per 1,000. It is true these men are adults, generally in the prime of life; but in the common lodging-houses and room tenements of Hexham their rate of death would be higher than the average, probably upwards of 30 per 1,000, or from seven to eight times the rate in Portland prison. Fever has been banished from well-regulated gaols; it is, however, to be found in full activity in the cottages and room tenements of the poor. A cottage or house isolated and in the country, if surrounded by excrementitous filth, may have fever as deadly as the worst parts of a town. There is no safety but in cleanliness and in pure air.

Analyses of Local Waters.

ANALYSES OF LOCAL WATERS.—*Samples of Water.*

No. of sample.	Name of source.	Date and Time of collection.		Temperature.		Remarks.
		Date.	Time.	Air.	Water.	
		1852.		Degrees.	Degrees.	
1	Watty's well	1 Oct.	11 A.M.	47	48	Day cloudy.
2	Letch well -	,,	11½ ,,	49	48	,,
3	Penny well -	11 Oct.	11½ ,,	52	46	,,
4	Market-pant	,,	,, ,,	52	46	,,
5	Tyne river -	,,	,, ,,	53	45	Sunshine.

These samples of water, from the several places as named, having been forwarded to London, were analysed, the results being as herein given.

The samples were numbered merely, the names of the sources not being given ; the report is as under :—

Report on Samples of Water from Hexham, by Mr. P. H. Holland.

" *To Robert Rawlinson, Esq., Superintending Inspector.*

" REPORT on Five Specimens of Water from Hexham, in Northumberland.

" These specimens were sent in five stone bottles securely corked, which were numbered 1 to 5 respectively. In no case had the water undergone any apparent decomposition.

" *No.* 1.—This specimen was bright and clear, (except a few white particles like very fine sand which soon subsided); it was pleasant to drink, and had neither colour, taste, or smell.

" Tests showed the presence of carbonates, sulphates, and chlorides, of lime and magnesia.

" The hardness was 13¾° of Clark's scale ; it softened by boiling without evaporation to 8°, and by the addition of lime water, to 6¾°.

" *No.* 2.—This specimen has similar sensible qualities to No. 1, except from containing a few visible animalcules.

" Tests showed the presence in it also of carbonates and chlorides, with lime and magnesia. The residuum left by evaporation burnt with a peaty smell, but it does not contain enough vegetable matter to be objectionable.

" The hardness was 18° ; it softened to 7° by boiling without evaporation, and to 7½° by precipitating the bi-carbonate of lime by lime water.

" *No.* 3.—This specimen also was bright, clear, and colourless, without taste or smell, though containing a small quantity of peaty matter.

" Tests showed in it the presence of carbonate and chlorides, with lime and magnesia.

Analyses of Local Waters. 39

" The hardness was $15\frac{1}{3}°$, after boiling without evaporation, $5\frac{1}{2}°$, and after adding lime water, $5\frac{1}{2}°$.

" *No.* 4.—This specimen also was bright, clear, and colourless, and free from taste and smell.

" Tests showed the presence of carbonates and chlorides of lime and magnesia.

" The hardness was 12°, after boiling without evaporation $8\frac{3}{4}°$, after adding lime water, 6°.

" *No.* 5.—This specimen was clear, but strongly tinted brown with peaty matter in solution. It had no perceptible taste or smell, but may probably have both in hot weather. Unless this water can be obtained more free from peaty infusion, the source is ineligible for a town supply.

" Tests showed the presence of chlorides and carbonates of lime, and a little magnesia.

" Its hardness was $6\frac{1}{2}°$; it did not soften by boiling a short time, or by the addition of lime water.

" The residuum left by evaporating at boiling heat was 11 grains per gallon, of which $4\frac{1}{2}$ grains burnt off with a strong peaty smell.

" The specimens, No. 1, 2, and 3, are nearly similar in quality, and may be thus arranged in degrees of eligibility, 1, 3 and 2.

" In sensible qualities the four first specimens are nearly alike. In No. 2 alone were any visible animalcules observed, but possibly the other specimens may contain them sometimes.

(Signed) " P. H. HOLLAND."
" 12th February 1853."

REMARKS.—In any new waterworks for Hexham, the water should be taken direct from the springs by covered conduits into covered tanks or reservoirs, and from thence by means of cast-iron pipes, direct into the town, and so by branch service pipes be delivered pure, bright, and cool,—spring water, free from animalcules or vegetation.

The economy of a full water supply may be shown by the following calculation:—If in a single tenement there is a use of 30 gallons per day, this will be 10,950 gallons per annum, equivalent to 109,500 lbs., or nearly 49 tons. This weight of water will be delivered at the tap inside the house, and on any floor, or at the top of the highest house. To carry this weight of water by four gallons, or 40 lbs. at a time, a distance of say 100 yards, would require $7\frac{1}{2}$ journeys each day, or $2,737\frac{1}{2}$ journeys per annum, a distance backwards and forwards of 547,500 yards, or 311 miles and 140 yards. Many of the poor inhabitants of Hexham would, however, think themselves well off if they could obtain good and wholesome pump or well water even by such an amount of labour. There is, however, another most important feature of economy in a well regulated general supply of water to a town, namely, the saving by softness. The spring water will be about 6 degrees of hardness; the water from local wells and pumps varies; but if 16 degrees are taken, the difference will stand as under:—

100 gallons of water at 6 degrees of hardness, will require 13 ounces of soap.
100 gallons of water at 16 degrees of hardness will require 34 ounces of soap.

Or the cost in soap will be three-fold that of the softer.

If there are 5,000 gallons of water required annually for purposes of washing by each cottager, the difference in cost will be as under:—

Water of 6 degrees of hardness will require 636 ounces of soap.
   „ 16   „   „   1,700   „

Comparative cost in soap.

	£	s.	d.
1,700 ounces, or 106 lbs., at 5d.	2	4	0
636 „ or 39¾ lbs.	0	16	7½
In favour of soft water	£1	7	4½

But besides the economy in soap, clothes are much more easily washed and are also much less worn by washing with soft water. There is also greater economy in tea, in brewing, and in all cooking operations.

These calculations will serve to show the owners and tenants of cottage property in Hexham, that a full supply of soft and pure water is a great pecuniary benefit to any neighbourhood. The advantages to the poor, as regards health and comfort, are beyond calculation.

PROPOSED WORKS.—The following estimates are given to show the relative cost for works to the ratepayers. Works of a similar character have been let and executed for less sums than are here set down, as the recent rise in labour and materials is allowed for.

Estimate for Water Supply.

	£	s.	d.
Collecting-mains from springs and covered reservoir	700	0	0
Mains and branches	1,500	0	0
Compensating reservoir	500	0	0
Valves and hydrants	300	0	0
	3,000	0	0
Add for contingencies, 10 per cent.	300	0	0
	£3,300	0	0

Estimate of Works.

Sewerage and Drainage.

	£	s.	d.
Main sewers and secondary branches	1,500	0	0
Street gullies and branches	300	0	0
Manure tank	200	0	0
	2,000	0	0
Add for contingencies, 10 per cent.	200	0	0
	£2,200	0	0

Abstract.

	£	s.	d.
Waterworks	3,300	0	0
Sewers, drains, and manure tank	2,200	0	0
	£5,500	0	0

Annual Expense.

5,500*l*., at 4 per cent., to repay principal and interest in 30 years, will require 318*l*. 1*s*. 8*d*.

Income from Water.

	£	s.	d.
667 inhabited houses, average rental for a full supply of pure water, inclusive of a bath, a watercloset, or soil-pan, say, 8*s*. 8*d*. per annum	289	0	8
Large consumers, hotels, &c., say	50	0	0
	339	0	8
Deduct annual payment, interest, &c.	318	1	8
Leaving in excess, towards management, &c.	20	19	0

N.B.—Any income from sewerage will be in addition to this sum.

ROAD FORMATION, AND STREET PAVEMENT.—There are many different opinions as to how streets and roads ought to be formed, as also what description of material makes the best carriageway. But the whole question is much too wide for due consideration in this report. There are, however, certain broad principles which may be stated:—

1st. The site of either street or road should be well drained, and in the suburban or country road open side ditches should be filled in; a foundation of dry materials should in all cases be laid to protect the subsoil from wear, or from working up to the surface. Most of the mud found on paved roads has, by vibration, risen up through the joints to the surface: a well-formed foundation prevents this.

2d. The cross section of a street or road should be almost flat, that the wear may be uniform over the whole surface. A proper arrangement of side channels will remove the surface water. A road too much rounded is worn at the "butt" or crown, whilst the sides have little or no traffic over them.

3d. A good pavement should have a smooth and even surface, that wheeled vehicles may pass over it at the usual velocities with the least jolting. The materials should be hard, to resist wear, and should offer sufficient footing to horses to ensure perfect safety.

No form of pavement or Macadam will wear with full effect unless the substratum upon which it rests has been made dry and solid. A double tier of pavement might be laid with singular advantage where there is excessive traffic. That is, the substratum may be paved with inferior material, and be fully rammed into solidity previous to laying down the true pavement. This would prevent that unevenness of surface, which does not proceed from actual wear of the surface materials, but arises from the yielding of the foundation.

It has been ascertained that a light and quick traffic displaces the pavement of a street more rapidly than heavy loads moving at comparatively slow rates.

Mr. Telford, in his Sixth Report to the Commissioners for the road from London to Holyhead, states his opinions on road-making as follows :—

" In order to ascertain the most effectual way of rendering the driving way hard and smooth, I caused an experiment to be made along a quarter of a mile, at the northern extremity of a road, by constructing the roadway with a bottoming of Parker's cement and gravel, and with a coat of Hartshill stone laid upon it ; and to ascertain what would be the comparative effect of using the same stone on the old surface of the road, I had a large quantity of it laid on between the Arch and the Holloway road. The result is, that between the months of October and March last, full 4 inches of the stone on the old road, between the Arch and the Holloway road, was worn away, where 8 inches had been laid on, while not 1 inch was worn down where it was laid on the cement bottoming. This result corresponds with other trials where bottoming has been made with rough stone pavement.

" This leads to the conclusion, that in all cases where the subsoil of a road is clay, (and where stone cannot be procured at any moderate expense), a bottoming of cement and gravel ought to be adopted.

" Pursuant to the authority given me by the Commissioners, a contract has been made for laying a cement bottoming 15 feet in width in the middle of the whole of this road, for coating it with Guernsey granite, and for covering the sides of the 15 feet with 4 inches of strong gravel, and for making footpaths. The complete success of the experiment I allude to at the northern end of the road, justifies me in expecting, that as soon as this mode of improving the whole is carried fully into execution, this hill, which has hitherto been so serious an obstacle, will be passed over by carriages without any difficulty, and that the road will be afterwards kept in repair at a much less expense than formerly.

" It is but justice to my assistant, Mr. M'Neil, to mention that this plan of forming a cement bottoming was suggested to me by him ; a plan which promises to be of the most important advantage in those districts of the country where stones cannot be procured at a moderate expense.

" The different parts of the Holyhead road which have been newly made with a strong bottoming of stone pavement, place beyond all question the advantage of this mode of construction ; the strength and hardness of the surface admit of carriages being drawn over it with the least possible distress to horses. The surface materials, by being on a dry bed, and not mixed with the subsoil, become perfectly fastened together in a solid mass, and receive no other injury by carriages passing over them than the mere perpendicular pressure of the wheels ; whereas when the materials lie on the soft subsoil, the earth that necessarily mixes with them is affected by wet and frost, the mass is always more or less loose, and the passing of carriages produces motion among all the pieces of stone ; which, causing their rubbing together, wears them on all sides, and hence the more rapid decay of them when thus laid on an unprepared soil, than when laid on a bottoming of rough stone pavement. As the materials wear out less rapidly on such a road, the expense of keeping it in repair is proportionally reduced. The expense of scraping and removing the drift is not only diminished, but with Hartshill stone, Guernsey granite, or other stone equally hard, is nearly altogether done away.

" Experience proves that there are no grounds for a common notion, that when materials are laid on a rough pavement, they are soon crushed by the wheels of carriages ; when the body of materials is six inches thick, no such effect is produced by the wheels."

ABSTRACT of the PUBLIC HEALTH ACT, with Remarks as to the Powers it contains.

" One objection, which had great weight with ratepayers and owners of property, is, that carrying out the provisions of the Act will cause a ruinous expense. This is proved not to be the case in those towns where the Act is in force and works are completed. Property in such towns has risen in value more than the cost of the works.

" By the 37th section, the General Board are to approve or disapprove of the *removal* of the surveyor, the power to appoint this officer remaining in the hands of the Local Board, without any reference to the General Board. The duties attaching to the office of surveyor are of a most important character ; and as a faithful discharge of them by the surveyor might not meet with the approval of some parties upon the condition of whose property he would have to report, it was thought desirable by the legislature that he should have due protection in the performance of his duty, and not be liable to discharge without a good and sufficient reason; this given, the sanction to his dismissal would follow as a matter of course.

"OFFICERS OF HEALTH.—Under the 40th section, the General Board are to approve or disapprove of the appointment or removal of the officer of health, and to prescribe the duties to be performed by him. The appointment of an officer of health is *quite optional* on the part of the Local Board.

"SURVEY.—By the 41st section, the General Board are to prescribe the scale of any survey made under the provisions of the Act. No plan of drainage can be efficiently and economically executed in the absence of a correct survey.

"PLEASURE GROUNDS.—The sanction of the General Board is necessary to the providing by the Local Board of public walks or pleasure-grounds (section 74).

"WATER SUPPLY.—By the 75th section, the approval of the General Board is necessary to any contract for the purchase, &c. or construction of waterworks; and they are also to certify as to the terms which may be reasonable at which water may be supplied by a waterworks company to the Local Board.

"BURIAL GROUNDS.—The 82d section gives a power to the General Board, upon a representation from the Local Board, to close any burial-ground dangerous to health; and their consent is also necessary to the formation of a new burial-ground.

"APPEAL TO GENERAL BOARD.—Provision is given, by which parties aggrieved by the decision of a Local Board on the following points, may appeal to the General Board, viz. :—

"With respect to trades of a noxious or offensive nature, newly established (section 64).

"With respect to the levels and widths of new streets (section 72).

"With respect to the charging or apportioning *private improvement expenses* upon owners or occupiers (section 120), in relation to the following purposes, namely :—

"The expense of constructing or laying down house drainage (section 49).

"The expense of constructing (section 51) or keeping in order drains, waterclosets, privies, or ashpits for private houses (section 54).

"The expense of cleaning or filling up offensive ditches or drains; and

"The expenses of repairing private streets, not yet adopted as highways.

"MORTGAGE, &c. OF RATES.—The principal power given to the General Board is by the 119th section, which provides that it shall not be lawful for the Local Board to mortgage the rates without the previous consent of the General Board.

"These powers are of great importance, inasmuch as their

object is the protection of owners and ratepayers against improper charges for inefficient works.

"There are in every town parties who have conflicting and rival interests. The interests of owners are opposed to those of occupiers; and again, those of owners and occupiers of the higher class of property are frequently opposed to those of owners and occupiers of the lower class. The latter have a strong inducement to obtain an exemption from the payment of rates, inasmuch as their property forms a very large proportion of the whole.

"Attempts to transfer burdens are not uncommon, and an instance occurred a few years since, in one of the metropolitan parishes, because there was no external authority to interfere and prevent such jobbery, in which a majority succeeded in making over to a minority a burden of upwards of 20,000*l.* per annum.

"Experience has shown that power given to Local Boards of mortgaging the rates, would, if not accompanied with some check, be attended with results the reverse of beneficial.

"The great object is to provide that the works shall be commensurate with the proposed expenditure; and that can only be achieved where there is a disinterested check on the expenditure.

"As a security for the protection of minorities and individual ratepayers, and of other interests involved, the General Board have stated that before they can give their sanction to the mortgage of rates for a long series of years, they must first be fully satisfied that the works for which money is required are of a permanent nature, and the advantages to be derived from their execution equal to the value of the improvement rate required to pay off the principal and interest; and, moreover, that they are not only efficient, but also economical.

"POWERS GRANTED TO LOCAL BOARDS.—Those powers may be divided into two classes, viz. :—

"Those which are *imperative,* and
"Those which are *permissive.*

"Those powers which it is *imperative* on a Local Board to carry out in all their integrity, are such as are absolutely necessary for the protection of the inhabitants, and were, on that account, rendered compulsory; as any delay or supineness on the part of a Local Board might be attended with fatal effects.

"The powers termed *permissive* are those which are left to the discretion of the Local Board to execute.

"DRAINAGE (POWERS FOR).—*Public Sewers, &c.*—By the 45th section, the Local Board are to repair all sewers vested in them, and to cause such sewers to be made as may be necessary for *effectually* draining their district; and such sewers are to be covered and kept so as not to be a nuisance, or injurious to health, and to be properly cleared, cleansed, and emptied. By the 58th section, all ponds, pools, open ditches, &c., and places containing any drainage, filth, water, &c., of an offensive nature, or likely to be prejudicial to health, are to be drained, cleansed, covered, or filled up by the Local Board.

"Other powers, of a minor character, are given in aid of those above mentioned.

"*House Drainage.*—However complete a system of main sewerage may be, it is of little use unless it receives the refuse from houses in the district In many Local Acts the powers for house drainage are very defective, leaving the works incomplete; consequently, the inhabitants do not derive the benefits contemplated, and which they had expected, while they have had to defray the whole cost of such imperfect works.

"By the 49th section of the Public Health Act it is provided, that it shall not be lawful newly to erect or rebuild any house, or to occupy any such house, unless and until a covered drain or drains be constructed, of such size and materials, and at such level and with such fall, as may appear to the Local Board to be necessary for the effectual drainage of the same; and such drain or drains are to communicate with one of the main sewers. Any infringement of that enactment renders the offender liable to a penalty not exceeding 50*l.*, together with full costs of suit.

"The same section also provides, that with regard to any house which may be without a proper drain or drains, communicating with a main sewer, the Local Board are to require the owner, or occupier, forthwith to construct or lay down such drain or drains as may be necessary, and communicating with one of the main sewers.

"If the owner or occupier of any house should not comply with the request of the Local Board to lay down such drains as may be necessary, the Local Board are empowered (by section 49) to construct such works as may be required, and to recover the expenses, either in a summary manner, or declare them to be private improvement expenses, to be repaid by a rate thrown over thirty years.

"WATER SUPPLY.—By the 75th section, the Local Board, where any house is without a proper supply of water, and such supply can be furnished at a rate not exceeding *twopence* per week, are to require the occupier to obtain such supply forthwith, and to do all such works as may be necessary for that purpose.

"If the request above mentioned be not complied with, the Local Board are empowered to do all the necessary works for affording such supply, and to levy a rate, not exceeding twopence per week, and to charge the expenses of the works as private improvement expenses.

"The Local Board may also provide their district with a proper supply of water, constantly and at high pressure; and for that purpose they may contract, purchase, or construct waterworks. In the event of any waterworks company being established in any district, the Local Board cannot construct waterworks so long as such company may be able and willing to afford a proper and sufficient supply of water upon reasonable terms.

"HOUSES, &c.—By the 47th, 53d, and 72d sections, full powers are given to the Local Board for regulating the erection of houses, &c. within the district.

" The Local Board are also required to see that houses are properly furnished with necessaries (*privies*), and that they are kept in order, for the decency and comfort of the inhabitants.

"STREETS, HIGHWAYS, &c.—Under the 117th section, the Local Board are to execute the office of surveyor of highways; and they are required (section 58) to cause the streets to be paved and repaired, and also properly cleansed and watered (section 55).

" The approval of the Local Board is also necessary to the laying out of a new street (section 72).

" The Local Board may also purchase premises for the improvement of streets (section 73), and may require owners or occupiers of private streets to sewer, level, and pave the same, and in their default, to do the necessary works and charge them with the expenses as for private improvements (section 69).

" Other powers are granted for the better regulation of streets, &c.

"LIGHTING.—Under the 8th section of the Public Health Supplemental Act, the Local Board are empowered to contract for the public lighting of the district.

"BURIAL GROUNDS.—*Interment of the Dead.*—Under the 81st section, the Local Board may provide reception-houses for the dead, and make arrangements for the decent and economical interment of any corpse received into any such house.

" The 82d section contains powers for closing any surcharged burial-grounds, and for the establishment of general cemeteries, for persons of all religious denominations.

"NUISANCES, &c.—Full powers are given for the removal of all kinds of nuisances, and also for prohibiting the establishment of noxious trades, &c.

"PUBLIC NECESSARIES (PRIVIES).—The Local Board may provide public necessaries (section 57).

"PUBLIC PLEASURE GROUNDS.—The Local Board may provide, or contribute to the support of, public walks or pleasure grounds (section 74).

"SLAUGHTER-HOUSES.—Under sections 61 and 62, full powers are given for the registration and regulation of slaughter-houses.

" The Local Board may provide premises for the purpose of being used as slaughter-houses (section 62).

"COMMON LODGING-HOUSES.—By the 66th section, powers are given for the registration and regulation of common lodging-houses.

" [Two Acts were passed in the session of 1851, one (the 14 & 15 Vict. cap. 28) for the well ordering of common lodging-houses, and the other (the 14 & 15 Vict. cap. 34) to encourage the establishment of lodging-houses for the labouring classes.]

"RATES.—1. *Special District Rate (sect.* 86).—This rate is to be levied to defray the expenses of making, enlarging, altering, or covering sewers (section 89), or of other public works of a *per-*

manent nature (section 89), and of other matters of a minor character, at the discretion of the Local Board.

" The payment of compensation for damages occasioned by the Local Board is charged upon this rate in certain cases. (*Vide* sections 54, 71, 80, 144.)

"*Property liable to be rated.* - Under the 89th section, the Local Board are empowered to subdivide their district as circumstances may require, and they are accordingly enabled to charge the expenses for which this rate is required to the property actually benefited by the outlay.

" The rate is to be levied on property assessed at the full net annual value ascertained by the poor rate (section 88).

" Land used as arable, meadow, or for pasture only, or as woodlands, market gardens, or land covered with water, or land used only as a canal, or towing-path for the same, or as a railway, to be assessed upon one fourth part only of such net annual value (section 88). Exemptions under any Local Act are not to be interfered with (section 88).

"*Persons chargeable.*—The occupier (under section 88) becomes chargeable to this rate, *except* in the case of premises under the annual value of ten pounds, or let by the week or month, or in separate apartments, at rents payable at periods shorter than quarterly; in which cases the Local Board may call upon the owner to *compound*, or (upon his refusal to *compound*) to pay the rate in full (section 95).

" This rate is not leviable whilst the premises are *unoccupied* (section 89).

" The rate may be spread over a period of thirty years, but must be so distributed as to pay off the expenses in respect of which the rate is made, together with interest at five per cent., within that time (section 86).

" The owner or occupier may, under section 92, redeem the property from the payment of the rate. The rate may be either prospective, or for expenses incurred within six months, retrospective (section 80).

" 2. *General District Rate* (*sect.* 87).— *Objects for which levied.* —This rate is to be levied to defray the expenses of the preliminary inquiry, the election of the Local Board, salaries, &c. of local officers and servants, and certain other casual expenses of executing the Act, which could not be defrayed out of any other rate, or out of the district fund account. (*See* section 87).

" The rate may extend over the whole or part of a district (sections 87--89).

" The district fund account will consist of the proceeds arising from the sale of sewage, &c., and of penalties recovered under the Act, together with other sums received by the Local Board.

"*Property liable to be rated.*—The rate is to be levied on property, with the same exceptions and exemptions as the special district rate—

" And is chargeable to the occupier or owner in the same manner.

"The rate is to be made as occasion may require, and may be prospective or retrospective, as in the case of the special district rate.

"3. *Private Improvement Rate (sect.* 90).—*Objects for which levied.*—This rate is to be levied to defray expenses incurred by the Local Board (upon default of owner and occupier), in respect of works of private improvement, viz., constructing or putting in order house-drains, waterclosets, &c.; draining, cleansing, covering, &c. offensive ditches upon private premises; putting in order private streets, and providing apparatus on the different premises for supplies of water for domestic use.

"*Premises liable to be rated.*—The rate will be levied on the premises in respect of which the expenses were incurred (section 90).

"The premises to be assessed in the same manner as for the other rates.

"*Persons chargeable.*—The parties chargeable are also the same.

"This rate may be spread over a period not exceeding thirty years, but must be so distributed as to pay off the expenses in respect of which the rate is made, together with interest at five per cent., within that period (section 90).

"4. *Water Rate (sect.* 93.)—*Object for which levied.*—For water supplied for the purposes of domestic use, cleanliness, and house drainage (section 93).

"*Property liable to be rated.*—To be levied on the premises supplied with water (section 93).

"*Persons chargeable.*—The parties chargeable are the same as for the other rates.

"The change annually made in the members of a Local Board precludes the possibility of their becoming the mere instruments of a Central Board, even if that board had any such power.

"The objections urged by those who insist that the Public Health Act is of a centralizing character are so opposed to facts, and so unsupported by sound argument, that it is difficult to believe they are seriously entertained.

"The Public Health Act, so far from taking from the ratepayers any powers which they have hitherto enjoyed, gives them a greater extension of power over their own affairs than they ever before possessed. The Public Health Act provides a compact and workable administrative body, in the election of which each ratepayer has a voice.

"A large number of towns are deterred from applying for Local Acts from the excessive cost, and the uncertainty of their obtaining them on a first application. A Local Act, giving but imperfectly the powers afforded by the Public Health Act for the combined purposes of drainage and water supply, will cost, if unopposed, not less than *one thousand pounds;* and if the bill is met with a parliamentary opposition, the cost will be more than *two thousand pounds.* The costs of the solicitor and parliamentary agent, in the case of the 'Bilston Improvement Act, 1850,' amounted to the sum of 3,463*l*. 9*s*. 5½*d*. This is not an uncommon case, as the expenses in other instances have much exceeded those for the Bilston Act. In contrast to which, it may be stated, that the cost of the application for the Public Health Act, on an average of seventy-seven towns, has scarcely been *one twentieth* part of the latter sum; and the Public Health Act provides that the payment of even that small amount may be extended over a period of five years, so as to press as lightly as possible on the ratepayers. On the other hand, there is a clause inserted in every Local Act passed, which provides that the expenses of obtaining such Act shall be defrayed *out of the first monies that shall come to hand*. The ratepayers are thus taxed to a large extent for obtaining what is frequently a very incomplete and imperfect measure, and, as experience too often shows, quite inadequate to meet the object its promoters had in view.

"A Local Act, while it contains a provision for raising money by the mortgage of the rates, very seldom contains any provision for paying off the money so raised; and the debt is consequently a perpetual burden on the inhabitants.

"The Public Health Act, on the contrary, provides that the money raised on mortgage of the rates shall be paid off by annual instalments, not exceeding *thirty;* so that the present ratepayers are only called upon to pay their fair quota of principal and interest for the first outlay, which will be distributed over a period of thirty years; and at the end of that time the town will be free from debt, with all the advantages existing for which the original outlay was made.

"The owners of small cottage property, and especially those who hold their property under leases for short terms, may have the whole outlay on their property thrown over a period of thirty years, so that the annual sum they will be called upon to pay will be but of trifling amount.

"There are many poor, industrious persons, who have been enabled, from the savings of many years, to purchase a single cottage. Such persons, although free from the payment of rent, have generally means barely sufficient for their maintenance; and to such, a payment, in one sum, of any private improvement expenses would be a virtual confiscation of their property. Those cases are anticipated by the provisions of the Public Health Act; and however small the amount expended on such property may be, there is a provision for its being thrown over a period of thirty years.

"Another favourable feature in the Public Health Act is, that a Local Board are empowered (by section 89) to divide their district into sub-districts, and to charge to each sub-district the actual sum expended for works within it. The object of this is, to provide against any person being called upon to pay for works from the execution of which he does not derive a benefit.

"Such are a few of the special advantages derivable from the Public Health Act."

SUMMARY.

CONCLUSIONS AND RECOMMENDATIONS.—I beg respectfully to present the following Conclusions and Recommendations for consideration :—

Conclusions and Recommendations. 51

Conclusions. 1. That Hexham is a "place" (a township) having a known and defined boundary.

2. That there are no Local Acts of Parliament which affect the township, the government being imperfectly carried on under the general Acts of the country.

3. That it is the wish of some of the principal landowners and ratepayers, as set forth in their petition, that the powers of the Public Health Act should be granted, that a Local Board may be formed for the several purposes contemplated and provided for in that Act.

4. That some of the ratepayers are opposed to the introduction of the Act, as see their petitions and memorials.

5. That epidemic, endemic, and contagious diseases prevail, and that the local mortality is excessive.

6. That the sewerage, drainage, and supply of water are at present very imperfect.

7. That the parish burial-ground is unduly crowded, and that there is no township burial-ground adequate to the present and growing requirements of the place.

8. That power is required to form, drain, and pave new roads and streets, to pave courts, to regulate slaughter-houses, and to remove all injurious refuse at short intervals.

9. That power to construct house-drains is required, as also to obtain a full supply of pure water, to the superseding of all local wells, pumps, and house cisterns. That the supply may be at constant service, high pressure, abundant, pure, and cheap.

Recommendations. Taking into consideration the ascertained deficient means for local government, as set forth in this Report and Summary of Conclusions, and the excess of mortality which prevails, I beg respectfully to recommend, that your Honourable Board will grant the prayer of the petitioning ratepayers, wishful to improve the town, and to lessen the local mortality, and will allow the Public Health Act, 1848, (except the section 50 in the copies of that Act as printed by Her Majesty's printers,) to be applied to the township of Hexham.

That the Local Board of Health to be elected under the said Public Health Act may consist of nine persons.

That every person shall, at the time of his election as member of the said Local Board, and so long as he shall continue in office by virtue of such election, be resident, as in the said Public Health Act, 1848, is required, and be seised and possessed of real or personal estate, or both, to the value or amount of not less than 1,000*l*.; or shall be so resident and rated to the relief of the poor of the township upon an annual value of not less than 30*l*.

That the elections take place for the whole of the district to which this Act is applied.

That the first election take place within one month after an Order in Council for the application of the Act to the township; and that the subsequent annual elections, to replace the retiring members, take place on the second Monday after the 25th day of March in each year.

I have the honour to be,
My Lords and Gentlemen,
Your most obedient servant,
ROBERT RAWLINSON,
Superintending Inspector.

To the General Board of Health,
    &c.   &c.

APPENDICES.

APPENDIX A.

REMARKS.—In directing the attention of your Honourable Board to the following petition and memorials, I beg most respectfully to state the following facts. The memorials against local improvement are, I consider, fallacious; the returns of the annual mortality show, beyond the possibility of dispute, that excess of disease prevails to the extent of nearly half of all who die, one of the consequences being heavy poor's rates. I can state, from personal inspection, that the water supply is deficient in quantity, impure in quality, and most defective in the mode of supply; sewerage proper there is none, and human refuse is heaped up behind and against dwelling-houses throughout the town; individual proprietors and tenants have no power to protect themselves against this form of nuisance, so fatal to health and life. The medical gentlemen in Hexham, who are active in striving to prevent improvement, cannot have made themselves acquainted with the opinions of the age and the returns of the Registrar General, which records the opinions of the Imperial Government, that imperfect sewerage, local nuisances, and excess of disease and death are cause and effect. An opposition, headed by medical men, is new to my experience; there must exist some private, and consequently, unworthy reason not apparent. After my inquiry and inspection, opposition, for the most part, ceased; the facts were too palpable and too strong to admit of present dispute, and, if the Act could at once have been put in force, it would have been assented to; but, with the necessary delay, has sprung up fresh opposition, which has been strengthened by mis-statements as to the cost of works, &c. The poor have no means of disproving the fallacious arguments used, neither can they appreciate benefits in perspective of which they have had no experience, and it is very easy to induce them to sign a memorial against works of improvement, which, in expense, they are told will be most oppressive, if not ruinous. It remains for your Honourable Board to protect these parties against their misleaders, and save them from the effects of their own want of knowledge.

PETITION FOR APPLICATION OF THE ACT.

WHEREAS by the Public Health Act, 1848, it is enacted, that from time to time after the passing of that Act, upon the petition of not less than one tenth of the inhabitants rated to the relief of the poor of any city, town, borough, parish, or place having a known or defined boundary, not being less than thirty in the whole, the General Board of Health may, if and when they shall think fit, direct a superintending inspector to visit such city, town, borough, parish, or place, and to make public inquiry, and to examine witnesses as to the sewerage, drainage, and supply of water, the state of the burial-grounds, the number and sanitary condition of the inhabitants, and as to any local Acts of Parliament in force within such city, town, borough, parish, or place for paving, lighting, cleansing, watching, regulating, supplying with water, or improving the same, or having relation to the purposes of that Act, also as to the natural drainage areas and the existing municipal, parochial, or other local boundaries, and the boundaries which may be most advantageously adopted for the purposes of that Act, and as to any other matters in respect whereof the said Board may desire to be informed, for the purpose of enabling them to judge of the propriety of reporting to Her Majesty, or making a provisional order as mentioned in the said Act ;

Now, therefore, we the undersigned, inhabitants of the township of Hexham, in the county of Northumberland, the same being a place having a known or defined boundary within the meaning of the said Public Health Act, and rated to the relief of the poor of and within that place, and being more than one tenth in number, and greatly exceeding thirty of the inhabitants rated to the relief of the poor of and within the said place, do hereby petition the General Board of Health to direct a superintending inspector to visit the said place, and to make inquiry and examination with respect thereto, with a view to the application of the said Act, according to the provisions of the said Act in that behalf.

Christian and Surname of each Petitioner.	Residence.
CHARLES HEAD,	Hackwood, near Hexham.
WILLIAM ROBB,	Hall Garth.
WILLIAM BELL,	Abbey, Hexham.

(And 149 others.)

PETITION AGAINST THE ACT.

To the Commissioners of the General Board of Health.

GENTLEMEN,

WE, the undersigned, being ratepayers and inhabitants of the township of Hexham, assembled at a public meeting, held at the Moot-hall on Monday, the 20th day of September instant, to ascertain the opinion of the ratepayers as to the necessity of " The Health of Towns Act" being extended to the town of Hexham, find that the greater portion of the ratepayers who signed the petition to the General Board of Health for the inquiry were not at the time aware of the nature of such petition, nor of the powers contained in the Act of Parliament ; and we feel satisfied that from the strenuous efforts now being made in opposition to the measure by the very parties signing that petition, had the nature of such petition been understood by them no such petition would have ever been presented.

We humbly and respectfully submit to the General Board our reasons for objecting to the measure.

There are sixteen public or open pants or fountains in the town, constantly supplied with more than a sufficient quantity of water, and together with the supply of water from private pumps and wells, is more than adequate to the requirements of the inhabitants; and there is no difficulty in any occupier having a sufficient supply of water on his own premises at a small cost.*

The town, from its peculiar situation, has always been, and still is, unquestionably, one of the most healthy towns in the kingdom; and the ratepayers feel perfectly satisfied that the averages for the last seven years, as applying exclusively to the township, are below the averages of deaths as required by the Act of Parliament.

The nuisances from time to time recurring, and any want of sanitary regulations, are amply provided against by the powers vested in the Board of Guardians and other local authorities, the nuisances being in themselves comparatively insignificant.

We further respectfully submit that the town of Hexham, situated as it is in an agricultural district, and depending on that district in a great measure for support, is not at present in a position to pay any further amount of taxation; and when we inform the Board that the present poor's rate is with great difficulty obtained, and is in fact, in many instances, in arrear, we feel perfectly satisfied that the Commissioners of the Board of Health will not at present send down Mr. Rawlinson, the feelings of the town being all but unanimous against the adoption of the Health of Towns Act.

(Signed) JOHN GIBSON.
ROBT. LYON.
(And 90 others.)

Priest-popple.
WILLIAM WEAR.
MARY RIDDLE.
BERNARD RICHARDSON.
(And 48 others.)

Hencotes.
JOHN JACKSON.
ANN FENWICK.
THOS. HEDLEY.
(And 32 others.)

Giles-gate.
JAMES MILES.
MATTHEW COULSON.
GEORGE PATTINSON.
ROBERT ROBSON.
JAMES DODD.
(And 67 others.)

Presented to the Superintending Inspector, by—
Mr. JOHN STOKOE.
CHRISTOPHER ANDERSON.
THOMAS BATY.
JAMES JAMIESON.
JOHN TAYLOR.

---

* The present water supplied is impure at its source, and it is quite impossible to give house supplies to the inhabitants from the present works.

Memorial against the Inquiry.

To R. Rawlinson, Esquire, Superintending Inspector, &c.

We, the undersigned, being ratepayers of the town of Hexham, assembled by public notice in the vestry room in Hexham, for the purpose of ascertaining the feelings of the town as to the necessity of adopting the Public Health Act, find that the feelings of the town and of the ratepayers are almost unanimous against the adoption of that measure.

The following are our reasons for refusing :—

There are no less than 16 public and open pants or fountains in the town, constantly supplied with more than a sufficient quantity of water ; and, together with the supply of water from private pumps, is more than adequate to the requirements of the inhabitants; and there is no difficulty in any occupier of the town having a sufficient supply of water on his own premises, if he chose, at a small cost.

The town of Hexham, from its peculiar situation, has always been, and still is, unquestionably, one of the most healthy towns in the kingdom ; and the ratepayers feel perfectly satisfied that the averages for the last seven years, as applying exclusively to the township, are below the average of deaths as required by the Act of Parliament.

The nuisances from time to time recurring in the town, and any want of sanitary regulations, are amply provided against by the powers vested in the Board of Guardians and other local authorities, the said nuisances being in themselves comparatively insignificant.

We further find, that many of the parties signing the petition presented to the General Board of Health were, by the propositions held out at the time, led to believe that the measure was one of a trifling nature.

(Signed)   Matt. Smith.
John Harbottle.
Wm. Lyon.
(And 6 others.)

Priest-popple.
William Wear.
Mary Riddle.
Bernard Richardson.
Eliz. Carr.
(And 53 others.)

Giles-gate.
David Lyon.
Robert Stokoe.
(And 186 others.)

Hencotes.
N. Wear.
Wm. Stokoe.
(And 46 others.)

Local Proceedings.

In reply to the statements made in these memorials against the application of the Public Health Act, I beg to direct attention to the following requisition and report. The present supply of water is not only limited, but it is at all times impure, and during rain totally unfit for use.

" *To the Bailiff.*

" WE respectfully request you to call a public meeting of the inhabitants to consider the best means of obtaining for the town a larger and purer supply of water.

(Signed) " WM. ROBB.
" WM. W. GIBSON.
(And 25 others.)

" First resolution, proposed by Mr. Robb, seconded by Mr. Pearson,—' That this meeting approve of the scheme just described and contained in Mr. R. Nicholson's Report, dated August 1849, and recommend immediate measures to be taken for the purpose of putting it into practical operation.'

" Mr. Head moves, Mr. Robb seconds,—' That a committee of 16 be appointed to consider the best means for carrying the above-mentioned object into operation, and report to a public meeting; and that such committee consist of Jasper Gibson, Charles Head, and 14 others.'—Carried unanimously."

In compliance with the above-written requisition, I call a public meeting of the inhabitants of Hexham for the purpose therein mentioned, to be held at the Moot-hall, on Monday the 4th day of August next, at one o'clock p.m.

(Signed) J. GIBSON, Bailiff.
Board Office, July 30, 1851.

WATER SUPPLY.

Introductory remarks by the Committee appointed to inquire into the supply of water, and to obtain professional assistance.

" *To the Inhabitants of Hexham.*

" OUR object in addressing you through the medium of this pamphlet is to draw your attention to the present supply of water to the town, and to lay before you some plans for its improvement. Few questions relating to the sanitary condition of a town occupy a more prominent position, at present, than that of an abundant supply of pure water. Its essential importance to health is now universally acknowledged, and recent experience has proved that it cannot be disregarded with impunity. But, important as this question is, it has not received that amount of attention in Hexham to which we believe it to be entitled. However deeply the necessity for a larger and purer supply of water may have been felt, no well-regulated efforts have ever been made to obviate it,

" The old pant-head system, which has been in existence upwards of one hundred years, has been allowed to continue unquestioned, although it has long been notoriously incapable of supplying the town with water in necessary quantity or quality. It is not probable that a system, which was doubtless suitable for the time when it was first established, will be equally well adapted to supply the increasing wants which such a lapse of time produces. The changes in our social and domestic life, the increase of population, and the new and enlarged views of the requirements of health, demand either a proportionate improvement of the old system, or the adoption of new methods better fitted to the wants of the age.

"We shall endeavour to prove the inadequacy of the present method to supply us with water in sufficient quantity, of good quality, and general distribution.

"1st. *Quantity.*—From the pant-head the town receives daily about 24,000 gallons of water. It is estimated that from the various wells and pumps a further supply of 12,000 gallons is obtained, making in all 36,000 gallons. Taking our population at something above 4,000, we have about eight gallons per day for every individual. This quantity is inferior to that which many towns give to their inhabitants. Nottingham gives 20 gallons a day; Liverpool 12 or 14; Exeter 18; Hartlepool 12; and many other towns in similar proportion. The best authorities on this subject hold that nothing less that 10 gallons daily per head will serve all the various domestic and other purposes which a town requires. But in Hexham a much less quantity than 8 gallons daily is used. From the present construction of some of the pants, a large per-centage of water escapes unused, and we cannot calculate the daily use of water for every individual at more than 5 gallons. It is true that the larger quantity actually enters the town, but all the conditions of sanitary improvement cannot be served unless the whole of it, at least, be used. This can only be done by affording greater facilities and inducements for its use. If arrangements were made to supply a large number of houses by service pipe at a trifling cost, all that could be supplied would be used, and the cleanliness and health of the town be greatly improved. We possess many natural advantages, of which we have not hitherto sufficiently availed ourselves. The quantity coming into the town could be increased at a comparatively small expense, and without undue interference with any vested right in the water-course. For the way in which this is to be accomplished, we refer you to the official report contained in this pamphlet.

"2d. *Quality.*—It is a fact too notorious to be denied, that the water from the pant-head, forming as it does two thirds of all the water entering the town, is impure in its best condition, and can only be prudently used for inferior domestic purposes. This will be no matter of surprise if the almost total want of arrangements to purify it be considered. Near the source of the stream from which the supply comes, there are brick and tile works, whose muddy refuse flows into the burn; it is exposed in its whole course to the cattle in the fields, and from the nature of the district through which it flows, must carry along with it a large quantity of vegetable matter in solution. When it reaches the pant-head this mixture, instead of being allowed time to subside and purify itself, is permitted to run, almost unchecked, into the town, and in consequence cannot, with due regard to health, be used as a common beverage. A few fir branches, interposed between the stream and the pipe, is all the purification it receives, and the other arrangements at the pant-head are so contrived as to give us all the surface water with its filthy scum, while the lower and purer part of the stream runs down to supply the tanneries and other works in its course. It cannot but be wondered at that a town pretending, and not without reason, to a considerable amount of intelligence, should have so long allowed its supply of an all-important necessary of life to be furnished by such rude, absurd, and vicious methods.

"Besides these general defects, no means being taken to pass aside a quantity or unusually impure water, a sudden shower of rain deprives the inhabitants of three fourths of the town of serviceable water for a day, and we leave it to every householder to consider during how many days in the year this is the case, and to how much inconvenience he is consequently put. To give some idea of the quantity of earthy and other matter held in solution by this pant-head water, we need only state, that in the course of a few weeks, a deposit of four inches of mud was formed in Gilligate-bank pant. For the means by which all this is to be remedied, we again call your attention to the annexed report.

"3d. *Its Distribution.* – This is the only point in which, of late years, any improvement has been attempted. Pants have been set down in various localities, and supply has thus been brought nearer to the consumers. Now, while granting this to be an improvement upon the whole, we cannot agree that public pants are the best and wisest modes of supplying a town with water. There are serious moral disadvantages resulting from them which it should be the common object to obviate. House service at high pressure is the best way by which we could get rid of all the evils attached to the present system, and is the only method of distribution which will be found really cheap and agreeable. Mr. Rawlinson, inspector under the Health of Towns Act, said at Newcastle, 'The most economical mode of supplying water to a town was by public waterworks, furnishing a distinct supply to each cottage and room.' It is the cheapest if we only consider the great saving of labour in carrying water from the public pant, and the more efficient performance of domestic duties. The convenience of always having water in the house, and not being required to seek it, often with delay and difficulty, is worth a small advance in the sum charged. The improvements which we suggest would cost to those who may adopt them only a trifle more than is paid under the present unregulated system. Those who might object to a very insignificant advance in return for most important benefits, would still be supplied as before ; but all who have a due regard for true economy, cleanliness, and health, would not hesitate to have the water brought into their houses by service pipe. The evidence of those who have already done so, and the incessant applications from others who at present cannot be supplied, prove that the public feeling of the town is altogether in favour of this valuable reform. Another advantage, to which we have not hitherto alluded, must not be passed over in silence. If the plan laid down in the report be acted on, we will have greater power to cope with fire than we have had, for in the case of all the fires which have lately occurred in the town, water could only be obtained with much labour and delay, and sometimes its supply was quite inadequate for the purpose. Were pipes laid down in the streets, and water supplied at high pressure, abundant means of extinguishing fire would be constantly at hand, and thus a feeling of greater confidence, in the event of such an alarming occurrence again happening, would be generally experienced.

" Some statement of the immediate occasion of your thus being addressed, is due to you. With a view to attempt some such reforms in our water supply as we have described, a new water committee was proposed and appointed. The old body had been many years in office without effecting any striking improvement, although doing the best they could with a bad system and small funds; on the new committee taking office, the subject of a general reform was broached, and the advice of an engineer proposed, as the best means of obtaining the necessary information. The committee, however, as a body, did not consider themselves justified in devoting any portion of their funds to such an object, and by a portion of them, a private subscription was commenced, which, liberally headed by the bailiff, J. Gibson, Esq., was warmly responded to by the leading gentlemen and tradesmen of the town. R. Nicholson, Esq., Engineer to the Whittle Dean Waterworks, was engaged to make the necessary survey, and report on the best means of giving us a pure and plentiful supply of water. This Report is now laid respectfully before you."

The following report is given, as it explains fully the present water supply, and sets forth the requirements of the place :—

ENGINEER'S REPORT on the best mode of supplying HEXHAM with pure WATER.

" Arcade, Newcastle-on-Tyne,
" GENTLEMEN, "August 1849.

" IN accordance with your request, I have examined the present system of supplying Hexham with water, for the purpose of considering whether it would be advisable to repair the existing pipes and works, or to alter the mode of supply, and adopt a scheme which, at a very small outlay of capital, would enable every householder to obtain a constant in-door supply to the highest rooms in their houses.

" The town contains about 4000 inhabitants, and with the exception of about ten or a dozen houses, (into the ground floor of which the water is conveyed,) the supply is obtained exclusively from public pants in the streets, and vicinity.

" Three of these pants are fed from ancient wells or springs in a dean belonging to Mr. Kirsopp, on the south-east side of the town, and the water is conveyed from thence in very old metal and lead pipes. In addition to the springs, a considerable extent of land drainage flows down the dean, and although the water is highly objectionable after rain, there is no arrangement by which it may be rejected. It passes through the pipes to the town in whatever state it may chance to be at the time.

"Another public pant is situate in the 'Seal,' close to the town. It contains two kinds of water, but owing to its low position, the water would only flow from it to a very small part of the town. There are likewise two other public pants in the streets, with limited supplies.

"The wells and land drainage from Mr. Kirsopp's dean, afford the best and most abundant supply, and its elevation is so great that the water would flow from it through a properly arranged system of piping, into the highest stories of all the houses in the town; but the present pipes are so old, and in such bad repair, that I would not advise any amount to be expended upon them; and I have likewise shown that the present indiscriminate mode of taking the water is highly objectionable.

"I therefore recommend that a small reservoir be constructed in the above dean, at the place where the water is now taken to supply the town, and the supply carried from thence through an entirely new metal main into the town. It will be necessary in forming the reservoir to have proper arrangements for rejecting and passing aside the water when discoloured and unfit for use, so that the quantity collected and conveyed to the town may be at all times clear and pellucid. The springs being always considerable, the reservoir need not, I think, contain more than about one month's estimated future consumption; but it should be so formed that it may be increased at any future period, if necessary; as it is known from experience that when a constant high service is afforded, the consumption soon exceeds any previous calculations.

"The proposed main should be laid along the line of the present road, and enter the town in 'Battle-hill,' and one supply main then laid to the south towards Temperley-place, and another to the north to the pant at the end of Coastley-row; and from the latter point a branch to be left through the Market-place as far as the water may be required. The extent of piping in the town will obviously depend upon the demand for the water; in the first place I would lay the pipes along the streets to all the pants now fed from Mr. Kirsopp's dean, which will afford the same out-door supply that exists at present, and give every householder on the line of pipe an opportunity of obtaining it within their houses. The remainder of the town may either be piped at once, or deferred until the number of applications for water fully warrant the outlay, which I confidently expect will be within a very short period.

"Although I do not propose to reduce the number of public pants, I ought to observe, that they are generally considered objectionable in other towns, as affording opportunities for servants and others to congregate; yet this objection will in some measure be removed, by every householder having the means of obtaining a constant in-door supply.

"Tanners and others who are now supplied from Mr. Kirsopp's dean will not be affected by my proposed plan.

"Fireplugs may be attached to the mains in the town, and as there will always be a considerable pressure on the pipes, it would afford a great security in case of fire; inasmuch as the water may be used direct from the mains, without the assistance of fire-engines. In a sanitary view also, an abundant supply of water is very important, particularly when it is constant, and at high pressure.

"In the absence of a proper survey and sections of the site of the proposed reservoir, and of the exact length of piping that will be required, I can only give an approximate estimate of the cost. But I am of opinion that the expense of making the reservoir, and of laying a new main from thence as far as the several pants before described, will not exceed 350*l.* ; and the additional cost of laying supply mains along the remainder of the streets will be about 100*l.*, credit being given for the value of the old lead and metal piping now lying in the ground, and which may be taken up and sold.

"The new main ought not to be less than four inches from the reservoir to the town, and the street piping three inches and two inches.

"It has been found in large towns where a constant supply of water has been afforded to the inhabitants, that as many as nine tenths of the houses have availed themselves of it, but if we calculate that *one fourth* only, or say 150 to 200 houses in Hexham will take the water, and this is a much smaller proportion than in any town with which I am acquainted, it will produce an ample return for the capital to be expended.

"I am gentlemen,
"Your very obedient servant,
"ROBERT NICHOLSON."

REMARKS.—In devising new works, I would recommend that the service reservoir should be covered ; that spring water alone should be taken into it ; that the main should be 6 inches in diameter ; that no street mains should be less than 3 inches in diameter ; and that the water should be delivered within the houses and room tenements. The public pants cannot be used with advantage for private house supplies.

It was found that no local power existed to carry out the works proposed, and a private Act of Parliament, if unopposed, would cost one fourth or one half the estimate for the works ; the following resolutions were therefore passed :—

"RESOLUTIONS at a public meeting held at Moot-hall, 15th August 1851.

"At a public meeting of the inhabitants of Hexham, called by the bailiff, to receive the report of the committee appointed at a meeting held on the 4th inst., to consider the best means for carrying into effect the resolution of that meeting, approving of a scheme submitted to it for obtaining a better supply of water to the town.

"The report of the committee having been read,—

"Resolved, on the motion of the Rev. J. Hudson, seconded by Mr. Dinning, that the report be receievd and adopted.

"Resolved, on the motion of Mr. Head, seconded by Mr. Nicholson, that a petition to the General Board of Health be prepared, signed, and forwarded, praying for the application of the Public Health Act to this town.

"Resolved, on the motion of James Kirsopp, Esq., seconded by Wm. Bell, Esq., that a committee, consisting of the gentlemen named in the annexed list, be appointed for carrying into effect the recommendations of the report.

" J. GIBSON, Chairman.
" JAMES ARMSTRONG.
" ISAAC BATY.
(And 22 others.)

I beg to direct the attention of your Honourable Board to the following report:—

"REPORT of the Committee appointed at a public meeting of the inhabitants of Hexham, held on the 4th of August 1851, for considering the best means of obtaining a better supply of WATER to the town.

" *To the Inhabitants of Hexham, in public Meeting assembled.*

" In accordance with the resolution passed at the public meeting, held 4th August, your committee proceeded to inquire as to the best means of procuring for the town a more copious and a purer supply of water. At that meeting two plans were mentioned by which the object could be accomplished, either by forming a private company, which would erect waterworks, or by applying for and obtaining the Public Health Act, which would give to the Local Board of Health, elected under that Act, the power to form waterworks, and to use other means for the effectual accomplishment of the object sought to be obtained.

" Your committee, after mature deliberation, and a careful consideration of all the arguments that could be adduced, unanimously decided in favour of recommending to this meeting, that the powers contained in the Public Health Act be applied for ; and in making such recommendation, thinks it would not be out of place to state a few of the reasons that have led it to this conclusion.

" The objections to a private company are, that it would be entirely without legal power to obtain the ground necessary for a reservoir, to procure way-leave for the laying down of pipes, or in any manner to interfere with the constant flow of water from the existing pants, and, therefore, might never be able to accomplish the purpose intended, without incurring the expense of obtaining a local Act of Parliament, which, with such limited means and in so small a town, is out of the question. It has moreover been found impolitic and disadvantageous for any town to render themselves dependent on a private company for the supply of so important an element of public comfort as pure water. And with such weighty objections against a private company, your committee was not able to discover in its favour one advantage that may not be more effectually obtained by having in operation the Public Health Act.

"In favour of obtaining this Act, your committee found that it gave full power to carry out to the greatest extent, and in all its bearings, the most economical and efficient scheme for securing pure water. It gives power to construct a reservoir in any place that may be thought most suitable; to lay down such pipes as may be required wherever it may be thought advisable; to regulate the supply of water to the existing pants as may be deemed most in accordance with the general good; and to use every possible precaution that the water supplied be perfectly pure. It places in the hands of the ratepayers of the town an extensive system of legal power, by which they can cheaply and effectually obtain for themselves all that they can wish for, as to the supply and character of the water. In fact it gives, for the expenditure of a few pounds, all the advantages which many towns have spent thousands to obtain by local Acts of Parliament; and it retains in the hands of the town the supplying of its own vital necessities, instead of making that supply contingent upon the will of a private company, which would, in all probability, more frequently consult its own rather than the public good.

"There is but one point which some may, perhaps, be prepared to advance as an objection to the recommendation of your committee. It may be asked how much money will be required to obtain the application of the Act, and what increase of rate will be necessary to continue it in operation. Your committee have to say in reply, that the application of the Act to the obtaining of water only, will prove to the town a profitable, rather than an expensive undertaking. Your committee had evidence laid before it from the Government Inspector's Report for Morpeth on the subject, of which they here give a summary. Proposed expenditure for Morpeth, 2,500*l*. Income from the supplying of water to houses 250*l*., from large consumers 100*l*., making a total revenue of 350*l*. Deduct from this 100*l*. for working expenses, and it leaves a balance of 250*l*., being a return of 10 per cent. upon the outlay, and which, large as it seems, is but a moiety of the profits reaped by private water companies in some towns.

"Now, from local circumstances, your committee, by adopting the data of the Government inspector, (which they may with safety do,) could make out a much stronger case in favour of Hexham, but they are willing to rest their case upon an equal amount of returns, and expect in so doing they meet all fears that can arise from this source.

"Your committee finds, that in order to obtain the application of the Public Health Act, a petition must be forwarded to the General Board of Health, asking it to send down an Inspector to investigate the condition of the town with reference to its supply of water, &c., and that such petition must have the signature of at least one tenth of the ratepayers, or it would be sufficient to prove, from the averages of the mortality of the town during the last seven years, that more than 23 in 1,000 die annually. Now it has been ascertained that the average mortality of this township during the prescribed period has been 25 in 1,000; the town could therefore on this plea obtain a commission of inquiry. But your committee recommends that a petition be signed and forwarded to the General Board of Health, asking for the application of the Public Health Act to Hexham.

" In the course of your committee's deliberations, many other matters were brought before its notice, which it deemed might without presumption be mentioned in its report as worthy the consideration of this public meeting. It was ascertained that the Public Health Act, if fully applied, will confer many other advantages besides that of a sufficient and pure supply of water. It will enable you to construct a complete system of sewerage, which is admitted by all to be much wanted, and for which the town is so admirably adapted. It will also enable you to make sufficient and suitable provisions for the removal of dust and soil from every house; to insist that every necessary means be used to preserve cleanliness both within and without the dwellings of the inhabitants; to see that even the houses occupied by the poorest families are properly, and at the same time economically, supplied with water and all requisite out-door conveniences. It will give full legal power for increasing the gratuitous supply of water if thought necessary, without its proceedings being liable to obstruction. It will enable you to provide means for the sufficient cleansing and sweeping of your streets. Under its provisions all slaughter-houses must be regulated and kept pure and clean; all common lodging-houses will be registered, and the daily outrages against morality and decency practised in those places put an end to by enforcing proper regulations respecting the number of lodgers and their accommodation. It will impose upon the Local Board of Health the duties now devolving upon the surveyors of highways, extending and completing their powers for the more effective discharge of those duties, especially with regard to neglected private roads and public walks. It also provides for the regulation of burial-grounds, besides affording many other facilities for the improvement of your town, which your committee consider it is unnecessary here to mention. Your committee is unanimously and strongly convinced that the health, morality, safety, and cleanliness of your town would be very greatly promoted if the Public Health Act was applied to it. And your committee can therefore earnestly recommend that the necessary steps be taken to apply for and obtain that Act.

" Dated the 14th day of August 1851.

J. GIBSON.
CHARLES HEAD.
R. E. RIDLEY.
WILLIAM W. GIBSON.
WILLIAM PEARSON.
THOMAS STAINTHORPE.
WILLIAM PRUDDAH.
THOMAS CLEMITSON.
HENRY HART."

APPENDIX B.

REMARKS.—The following reports recapitulate some of the information which has been previously given in this Report; but, being a description in detail, by gentlemen resident in the town and neighbourhood, I have thought it proper to insert the reports entire, in the order they were supplied to me during my inquiry.

LOCAL NOTICE.

" PUBLIC HEALTH ACT, 1848.

" In July or August 1851, a requisition was presented to the bailiff to call a public meeting to consider the means of obtaining for the town a better supply of water, and he fixed a meeting for the 4th August.

" On the 4th August a public meeting was accordingly held at the Moot-hall, and a resolution passed, recommending immediate measures for carrying out a scheme contained in a report of Mr. Robert Nicholson's, in 1849 ; and a committee of 15 was appointed for carrying that object into operation.

" On the 12th August the committee met, and resolved, that in their opinion the best means of carrying the resolution into effect was by the adoption of the Public Health Act, 1848, and requested Mr. R. E. Ridley to prepare a report to be submitted to an adjourned meeting of the committee on the 14th.

" On the 14th a report was read by Mr. Ridley and adopted by the committee.

" On the 15th a public meeting was held at the Moot-hall, when the report of the committee was read and adopted; and it was resolved that a petition to the General Board of Health should be prepared, signed, and forwarded, praying for the application of the Public Health Act to this town, and a committee was appointed to carry into effect the recommendations of the report.

" 200 copies of the report were printed and circulated among the principal inhabitants, and the committee soon obtained the requisite signatures to the petition, which was then forwarded to the General Board of Health.

" The committee have since divided themselves into sub-committees for inquiring into and reporting upon the different branches of the subject ; and the following are copies of the reports of the sub-committees, the several members of which will attend the inspector to substantiate the statements therein contained. The committee will also produce a map of the town."

" REPORT OF THE WATER COMMITTEE.

" The town of Hexham would seem, at first sight, to be favourably situated for obtaining an abundant supply of water. It is placed near the foot of a range of hills of considerable height, and is traversed by three streams, descending from the high ground to the south and south-west of the town. There is, besides, a number of springs or wells in the town itself, which afford a partial supply; and yet, in spite of many natural advantages, it cannot in any sense be said that Hexham is well supplied with *pure* water. Of the three streams which run through it, one is too insignificant to be thought of as a source of supply; another, though much larger in volume, is, when it approaches the town, at such a level as must preclude its being made of much service, and but one is available for the purposes of bringing this prime necessary of life in any quantity to the town. Its capabilities have been long known, and in some degree appreciated and used. Nothing certain is known as to the time when the humble waterworks, conveying water from it to the town, were made; but there is sufficient evidence to show that they are very ancient, and must have had their origin in a time when the population of the town was greatly less than it now is, and when the importance of water, in its relation to cleanliness and health, was not so fully recognized.

"The stream is made up of several excellent springs, and the land drainage of a small district to the south of the town. It descends with considerable rapidity, and for a great part of its course runs through a thickly-wooded dell or 'dean,' with high steep banks. About half a mile from the town, and at an elevation above the Market-place of about 70 feet, advantage has been taken of a point where the banks draw near each other, to throw across a small dam or weir, about four feet high. The idea would appear to have been to obtain a small reservoir; but if this was its original condition, it has long been filled up by the gravel and debris brought down by the stream, whose bed is now on a level with the head of the dam. The water is now simply turned aside from its course a little way above this point, and made to run in a narrow channel towards that part of the dam from which it must proceed to the town. The only attempt at purification is by causing the water to flow through a few interwoven branches before it descends into a small covered cistern, from which a 3-inch pipe runs in the direction of the town. Its course lies along the bank of the 'dean,' until it reaches a cistern immediately behind the Catholic chapel, into which it empties itself. In this cistern are two pipes to carry the water to different parts of the town; but these pipes being of too small a bore to carry off all the water coming into the cistern, a considerable quantity is here allowed to run to waste. Of the pipes referred to, one enters the town by way of the Battle-hill, and descends to what is called the Cattle-market pant, which has an open cistern; thence a branch pipe is taken down Priest-popple, and near the foot of that street supplies a close pant, working by valve. The other pipe, after leaving the cistern behind the Catholic chapel, crosses the street near Mr. Singleton's. It gives off a branch to a small pant near the Scotch church, passes through the park and gardens belonging to the abbey, and appears at the market pant, the most important and ancient of all the pants in the town. The inscription upon it shows the present erection to have stood about 150 years, but there can be no doubt that, long anterior to that period, water was supplied to the Market-place from the same source, and by similarly primitive means to those which now exist. It appears to have been erected by the benevolence of a private gentleman. It is a large and open cistern, but judging from the state in which its neighbourhood too often is to be found, these properties of size and openness do not contribute greatly to cleanliness. From this point pipes depart in two directions; one proceeds through the

Market-place, Market-street, and the upper part of Gilligate, to a pant on Gilligate-bank. Another pipe descends to nearly the middle of the Bull-bank, to another pant, which, though peculiar in its construction, is like all the later erections, covered, and works by valve.

"This short and rapid sketch may convey some slight idea of the nature and extent of the only formal attempt ever yet made to supply Hexham with water. It embraces within its limits three fourths of the whole town, the other fourth being supplied by wells, to which further allusion will be made. The system has arisen without unity or regularity of plan. The primitive and inefficient nature of the arrangements at the pant-head has vitiated all that has since been accomplished. Addition after addition has been made in the shape of public pants, as necessity seemed to indicate; but the idea of affording house service does not appear to have been entertained. Within a few late years, indeed, since the advantage of this mode of supply began to be appreciated, between 20 and 30 houses have taken it in by pipe. But because the pressure is only taken from the level of the market pant, it cannot in any case rise higher than the ground floors, and even then the supply is of the most unsatisfactory description. This cause alone has prevented the rapid extension of this mode of supply which the Hexham public earnestly desire, and will account for the small proportion of house occupiers who have the privilege. The number of inhabited houses in Hexham, by the last census, was found to be 531, and the number of separate occupiers 1,043; and yet, out of this number, not more than three per cent. have the water carried by pipe into their houses, and even they have it in so inefficient a manner as greatly to neutralize the advantages which would otherwise accrue. The scale of charges for this attempt at supply has hitherto been arbitrary and irregular; some have been charged 5*l.* for leave to take a pipe into their houses, and have been relieved of any annual charge; others who have paid the above sum have also had to pay a small yearly rate, while a third class have not only escaped the payment of the premium, but have only paid such an annual sum as they pleased, often ridiculously trifling. As might be expected, from the fact that most part of the house service has no greater pressure than from the pants, this part of the system is not capable of great extension. The following are the only streets where there is house-service to the extent named; viz.—

Priest-popple.—A brewery supplied from Cattle-market pant;

Battle-hill.—A house from the pipe between Catholic chapel cistern and Battle-hill, and Mr. Singleton's house.

Back street.—Soup kitchen, Thomas Hedley, and Mrs. Fenwick's houses, from pipe in Abbey garden.

Market-place.—A house from Market-place pant.

Market-street.—Several houses from same source.

Gilligate. - Dye-works, brewhouse, &c., from Gilligate-bank.

Abbey.—From pipes in garden.

"In none of these cases does the water rise above the ground floor, nor can it from the want of pressure. The Church-row,

Defective State of the Water Supply.

Back-row, Bull-bank, Hall-garth, Fore-street, Skinners-burn, Hencotes, Gilligate, and Cockshaw, are totally unsupplied by house service; indeed all the town, except the few houses mentioned, and nearly all of them are in the same dilemma with regard to the level. These streets, on account of their slight difference of level from the source of pressure, cannot have the water brought into their houses, and however anxious their inhabitants may be for a better supply, they are forced to be content with the old, laborious, and tedious mode of carrying it from the public pants. But there is another circumstance which would even render any extension of house service not very desirable. This is to be found in the frequent state of impurity of the water. Though we have no analysis of the pant-head water, yet it is known to be soft, and long experience has proved it serviceable for general domestic purposes. But from the notorious want of care for its purification at the dam it is no exaggeration to say, that during a considerable portion of the year it is thoroughly useless even for the commonest purposes. There are several special reasons too, which demand special attempts to obtain purity. Lately brick and tile works have been commenced near the source of the stream, into which their muddy water easily flows, and so completely destroys its usefulness that the whole of the town dependant upon it have often from this cause alone been compelled to seek water from distant wells. The same result is produced whenever a heavy shower falls upon the high grounds near the head of the burn, because no effort has been made at the dam to turn aside the turbid water. Again, when none of these causes are operating, the very nature of the district through which the stream flows would prove that careful and constant purification is required. As has already been stated, its course lies through a thickly wooded and luxuriant dell. The quantity of decaying vegetable matter falling into the stream is considerable, and it cannot but be a matter of some consequence to get rid of this copious solution of deleterious substances.

"The quantity of water coming daily into the town, next demands our attention. It was ascertained by actual measurement that the quantity coming into the cistern behind the Catholic chapel amounts to 24,000 gallons per day, but on account of the waste continually going on at this point as we have already stated, nothing like this quantity enters the town. The sum of the bores of the two pipes bringing it to the town is greatly below the size of the pipe leading into the cistern. From all other sources it was not thought possible to obtain more than 12,000 gallons, and in this case too a considerable margin has been allowed. Now, at the last census in 1841, the population of the town was 4,601. It will, therefore, be found that, taking the supply in the most favourable light at 36,000 gallons daily, the allowance for each individual will be somewhat less than eight gallons. But there is nothing like this quantity actually used; two of the pants are open, and run continually to waste through at least half of the whole 24 hours. As to the probability of increasing this supply from the pant-head, it may be stated that about as much water flows down the regular channel of the stream as enters the town by pipe.

"The extent of main and service piping to and in the town is about two miles. The most of it is iron; from the pant-head to the two first pants its condition is such, that any increase in the quantity of water would necessitate an entirely new set of pipes. There can be no question, but that they have been in use from time immemorial, and as defects have been discovered, they have been remedied often by the readiest means and in the most insufficient manner. The pipes leading from the Cattle-market and Market-place to the branch pants, having been only laid down for a comparatively short period, are, as may be expected, in a much better state, but their bore would be insufficient to convey the quantity of water necessary for extensive house service.

"Having thus endeavoured to give something like a sketch of the pant-head system, some allusion must be made to the other sources of supply. The first we would mention is the Penny-well, in the Skinners burn-street. Its supply is scanty, but its water is held in some estimation for many purposes; the demand for it is greatly increased when the pant water is muddy, great numbers congregate around it, and have to wait sometimes for hours together. The Seal-well is near Cockshaw, on the west side of the town, and forms the principal means of supply for the numerous population of Gilligate. Its water is harder than any of those hitherto described, and is mostly in request as a drinking water; from its deficiency of elevation it cannot be made so useful as it might otherwise be. Towards Tyne-green, the distance to the Seal-well becomes too great to be travelled more than is absolutely necessary, and for all common purposes the inhabitants are indebted to the water of the river. There is also a small pant in Hencoats supplied by a spring; its cistern is but small, and quite inadequate to supply the street with its proper quantity, but in spite of numerous complaints no efforts have been made to increase its amount; besides these there are two or three wells of too small account to deserve special notice.

" Some of the pants are rather distant from each other, and from parts of the district they are intended to serve. We will instance the case of Dean-street. There is a numerous and crowded population, who must either come up to Priest-popple pant, or obtain their water from such private sources as may be accessible. Again, when the pant-water happens to be impure the inhabitants of all the following streets must go either to the Penny-well or Seal-well,—Market-place, Hall-gate, Bull-bank, Fore-street, Back-street, Priest-popple, Battle-hill, Gilligate-bank, &c. In fact, the whole town is then thrown upon these two pants and the few other insignificant wells scattered over it.

"One of the best proofs of the inefficiency of our present system is in the difficulty felt in contending against fire. It may be said that in almost every instance of late, the short supply of water has greatly magnified the danger and consequent alarm. We will, however, only quote the two most recent cases. The Blue Bell stables and haylofts, situated in close proximity to valuable property, and to warehouses containing tar, oil, and other combustible articles, were destroyed by fire; by the time the flames were subdued the water was quite exhausted. Had the fire communicated to the houses in the Fore-street or the warehouses on the opposite side of the yard the result would have been most disastrous, as the water had been drained from all the pants in the neighbourhood. In the case of Mr. Dodd's and Mr. Ord's property in the Hallgate, the supply of water from the regular sources had been used, and had it not been for the kindness of some gentlemen in lending carts and horses to bring water from the railway station, it is impossible to say where or when the fire might have ended. In both cases no more water could be obtained from the pants than sufficed to supply the fire-engine for one hour.

"All matters relating to water service in Hexham have been for many years under the management of a committee. For a long while this body was self-elected; collected and expended the money raised by voluntary contribution without ever having their accounts audited or published. They effected many valuable improvements, but their mode of conducting the business was by-and-by considered so unsatisfactory, that a general desire was felt and expressed to have some change in the constitution and operations of the committee. A few years ago, the ratepayers in vestry meeting appointed the committee; although the power to make this appointment might fairly be questioned, yet the new arrangement was acquiesced in, and the accounts were submitted yearly to the vestry meeting, and a new committee appointed. It was soon perceived, however, that the old system was not susceptible of much improvement, even from this new governing body. The impression that a total change was necessary gradually arose in the public mind, and this being strengthened by several incidental causes, has resulted in this effort to have the Public Health Act applied to Hexham.

(Signed)   WILLIAM ROBB.
               WILLIAM PRUDDAH.
               W. A. TEMPERLEY."

" SEWERAGE AND DRAINAGE REPORT.

" *To the General Committee appointed to put in operation, in the town of Hexham, the Health of Towns Act.*

" GENTLEMEN,

" YOUR sub-committee, under the Health of Towns Act, have, in compliance with your wishes, examined the present drainage of your town, and cannot refrain from expressing their opinion that no town can be found where the drainage is more imperfect, and few where the advantages are greater to an effectual drainage.

" Your committee commenced their survey at the head of Hencotes, and the following are the present drains :—

" 1st. *Hencotes.*—A small and imperfect drain is carried from below the house called the " Three Tuns," across the Hexham turnpike road, and passing along the west side of that road, empties itself into the Hencotes-burn on the west side of the bridge across that burn. Another commencing at the Hencoats-pant, near the almshouse, and passing on the south side of the street to the Scotch chapel, empties itself into the Seal-burn on the north side of the bridge.

"2d. *Battle-hill.*—A drain commencing on the south side of Mr. Kirsopp's garden, up Maiden-lane, and passing down that lane to the corner of Watson's house, and thence on the south side of the street, into the Catholic chapel grounds, where it empties itself into the Seal-burn on the south side of the bridge. Another commencing on the south side of the Battle-hill, near to Mr. Jasper Gibson's gates, and passing down on the same side of the street, empties itself into the Skinners-burn at Woodmas's house corner. Another on the north side of Battle-hill, commencing at the Golden Lion, and passing on the same side of the street, to the pant at Bow-bridge, when it empties itself into the Hall-orchard-burn.

"3d. *Priest-popple.*—A drain commencing at the house of Miss Moncrieff, and passing down the street until it reaches the path-foot, when it crosses the road and enters the close belonging to Mr. Thomas Jefferson, and then, by an open cut, is taken down and empties itself into the river Tyne, at the foot of Tyne-mills-close.

"4th. *Market-place.*—A drain commencing at the pant in the Market-place, and passing the corner of the shambles to the north side of the Fore-street, and so along the same to Barrett's house corner, where it empties itself into the Hall-orchard-burn. Another commencing in the Long-yard, and so passing on the south side of the shambles, empties itself into the drain along the Fore-street, at the corner of Peter Bredwell's shop. Another commencing opposite Mrs. Fenwick's property in the Back-street, and so down the Meal-market, to the corner of Bell the draper's, where it empties itself into the Fore-street drain.

"5th. *Black-bull-bank.*—A drain commencing in the Back-row, opposite the Police-station, and so on past the Post-office, down the Black-bull-bank as far as Murray's property, when the same empties itself into the open channel.

"6th. *Hall-garth.*—A small drain at the foot of the Moot-hall-stairs, passing through under the house of Mr. Thomas Jameson, empties itself into the open channel on the Black-bull-bank. Another, commencing near the property of Mr. George Bell, passes the property of Christopher Anderson, and empties itself into the Hall-orchard-burn, behind the property of Mrs. Grant.

"7th. *Gilligate.*—A drain commences at the west end of the Church-row, and is brought to the end of the Church-flags, past the property of Mr. Joseph Fairless, and crossing the Market-place, empties itself into the drain at the Market-pant. Another commences opposite the property of Richard Mews, and so down that part of Gilligate to the Haugh-lane-end, when it empties into the Gilligate-burn.

"The drains generally are built of rubble stone without lime, and flat at the bottom, and from their imperfect state, and their construction, are, in the opinion of your committee, better calculated as a deposit than for carrying away the sewerage of the town. It will be noticed that the greatest portion of the Back-street, the Black-bull-bank, and Gilligate, also Priest-popple, are without any drainage. Although the burn, passing down Skinners-burn, and so down the Hall-orchard, is now partly arched over, the arch is so imperfectly constructed that it has been repeatedly choked up, and has been attended with great inconvenience and damage to the inhabitants of that locality.

"The drains are all marked on the plan that accompanies this report, and shown thereon by green dotted lines.

"As regards the proposed drainage, your committee find that many private interests will have to be consulted as to the course of the drains through their property, and so many private rights to be considered regarding the deposit of sewerage, they deem it better to defer for the present any report upon them.

(Signed)   CHAS. HEAD.
D. W. ROME.
JACOB BULMAN.
JAMES ARMSTRONG.
"*Hexham, 23d February* 1852."   IS. BATY."

"Your committee see no reason to alter anything stated by them in the foregoing report, but would particularly call your attention to the nuisance produced by Mr. Thomas Pratt having brought a soil-pipe into the drain passing out of the Back-street down the Meal-market, and so into the drain along the Fore-street.

(Signed)   CHAS. HEAD.
IS. BATY.
D. W. ROME."

" SANITARY REPORT.

" The town of Hexham is pleasantly situated on the right bank of the river Tyne. It is built upon the lowest part of a hill, which is somewhat amphitheatre-like, and rises to a considerable height on the south. Between the town and the river extends a rich alluvial soil, clustering with fruit trees, which, with the adjoining hills covered with bright green meadows and stately woods, give to the neighbourhood an appearance of luxuriance and fertility. In a sanitary point of view Hexham must be considered as well situated. Although it is, from its proximity to the sea both east and west, much subject to very sudden changes of weather, to frequent rains, and to heavy fogs, yet the direction of the valley from west to east is favourable to constant currents of air, which are always fresh and unpolluted. For obtaining an excellent supply of water is is abundantly favoured. Upon the hill behind the town is found a sufficient supply, while the height and nearness of the hill afford cheap and easy means for making the supply copious and useful ; and as the town rises throughout its entire length, its sewerage works may be constructed without difficulty and at small expense, while the sewers, from the inclination they must necessarily have, will always be effective and clean.

"Hexham, although possessing every natural advantage for cleanliness and healthfulness, has, by its construction, been rendered comparatively filthy and unhealthy. Like all old towns the houses are much crowded in the centre of the town, so much so as to prevent the inhabitants from having ordinary conveniences, or if they have them, to make their existence almost as great an evil, if not greater, than their entire absence.

"The whole of the houses that enclose the Market-place are thus situated; on the west and north no conveniences at all exist, not one inch of ground to use as a yard. In a few houses are wooden erections which are used as privies by the families, one or two having the offal removed daily, others allowing it to accumulate until the whole neighbourhood is infected by the poisonous effluvia.

"In one instance the entire refuse of a family is deposited in a cellar immediately beneath the post-office, where it remains for months, the stench arising from it having an outlet through an iron grating, above which is the posting-box and delivery window of the post-office, so that the whole town is brought within the range of its infectious influence.*

"On the south and east sides of the Market-place are to be found, in the generality of cases, small yards, containing an area 5 or 6 feet square, and having ashpit and privy. On all sides the yards are surrounded by high walls, so that however much the wind may blow without, within the contaminated precincts no breeze ever comes. Windows in many instances overlook these yards, which when opened must necessarily admit the accumulated gases that are generated beneath, with results that cannot be but injurious.

"Close to the Market-place, extending in a southerly direction, is a range of houses dividing the two principal streets in the town, and having a frontage into each. These houses are placed upon the smallest possible compass of ground, and have connected with each, one of the unique yards we have mentioned. The stench from those yards at all times, but especially in summer, is represented as very offensive. The conveniences are necessarily between the dwelling-houses, so that in whatever direction the wind may blow the effluvia is carried into the dwellings on one side or the other.

"The ashpits in most cases are also the receptacles of the refuse water from the houses, and consequently from the walls is constantly oozing a putrid semi-liquid, which slowly spreads over the yards and down the channels, increasing greatly the evaporating surface, and thereby augmenting the evil. The houses in the Fore-street are much exposed to this form of nuisance, about which the inhabitants justly complain. And, as if the sanitary condition of the locality was not thus bad enough, pigs and horses are in some instances crowded into the yards.

"At the south-west corner of the Market-place debouches a passage which is the entrance to a large yard covered with the most active sources of nuisance. Dwellings partly surround the

---

* This has lately been discontinued.

square, one side being occupied by that portion of the parish church which is used for religious services. There are four slaughter-houses, with their accompanying offal-heaps, always in a state of active decomposition, also about a dozen piggeries, most of which have separate dung-heaps. A large stagnant drain exists, into which filthy water and vegetable and animal refuse is regularly thrown, while a large ashpit and two privies occupy the centre, where the refuse of some 15 or 20 families is daily deposited, and allowed to accumulate for months. At the end of the Fore-street is a yard in many respects similar to that just described, being scarcely so large but much more filthy.

" Towards the north-west from the Market-place extends a range of buildings fronting Market-street. The buildings are densely populated, and are provided with conveniences as follows :—Two privies constructed within the dwellings, emptied every morning, one in a yard barely sufficient to contain it; this is worse than the others, being so much seldomer cleaned. In a yard 12 or 14 feet square is a piggery, privy, and ashpit, which are cleaned but once or twice a year. There are two privies constructed on the public road, but kept private property. The remainder of the population, consisting of 19-20ths of the whole, have no means of getting rid of their refuse except that of carrying it to a considerable distance, which of course is seldom done. It is generally thrown into the open channel or into the churchyard. All are alike un-provided with the means of removing the refuse water. For their convenience in this respect a drain was constructed a few years ago, with an open gutter above it, but the ingenious mechanic to whom the work was entrusted has formed the drain and gutter so that the water actually runs in the reverse direction to that which it ought, and in consequence they are both at all times full, and are thus rendered serious nuisances instead of public benefits ; in addition, within three yards of the back doors and windows of these houses, is a very much overcrowded churchyard, the evil influence of which it is difficult to calculate.

" Further on in the same direction we reach Gilligate, where exist the same sanitary deficiencies, made more apparent and noxious by the less cleanly habits of the people. There are several groups of houses thickly inhabited, possessing no conveniences of any kind, and in a majority of instances where such exist, they are used by several families, and no one attempts to keep them clean, while the want of drainage causes stagnant pools almost constantly to exist. In one instance the drainage from a piggery, privy, and ashpit, has found its way into the room of an adjoining house, and in consequence of the yard being level with the second floor of the house, the liquid oozes through the wall, runs down from the second to the first floor, in such quantity as to wet the beds, and to require that a well should be sunk in the room several feet deep, in which these pestiferous drainings are allowed to accumulate until it is full, and then they are removed to make room for more. Throughout the whole of this part of the town, the want of drainage is very manifest. Its low position prevents it from having much advantage from natural drainage, so that in every back-yard exists stagnant pools for which no outlet can be had.

"To the north-east from the Market-place is Black-bull-bank, which unlike Gilligate possesses every facility for natural drainage, and in consequence, five drains open on to it, and their contents flowing down, form a stagnant pool at the foot. These drains are all of an offensive character, and are loudly complained against by the inhabitants. The conveniences for this district are in many instances very insufficient, several of the families being in this respect entirely destitute. There is no other means of removing the refuse water of the various families except by the open channel, which greatly contributes to augment the evils already mentioned.

"One house is provided with an ashpit under the dwelling-room, and although it receives the contents of a privy, it is not emptied more than once a year.

"Such are a few of the more glaring evils that are to be found in the sanitary condition of Hexham. Many more might be mentioned almost as bad, forming a gradation to the few houses that are better provided with conveniences. These are indeed very few, and in almost every case are far from being such as they ought. There are about some half dozen waterclosets to be found in Hexham, but in consequence of the absence of drainage, they prove serious nuisances to the neighbourhoods in which they exist. The structures intended for drains are, except in very wet weather, reservoirs, so that almost constantly, from every opening, but especially from those in the vicinity of a watercloset, exhale very offensive and injurious effluvia.

"Considered as a whole, the sanitary condition of Hexham is of the lowest class. The few attempts that have been made to improve it, have been too paltry and unconnected to be of much use; they have been made for individual rather than public good, hence their inutility, and in many instances ultimate injuriousness.

"It is not necessary to prove the consequences of such a sanitary condition upon the health of the inhabitants. Long courses of experience and observation by the best members of the medical profession have indisputably shown that in proportion as you have inferior sanitary regulations, you have causes of disease. It would therefore be quite sufficient to enumerate as above the evils of this kind that exist.

"It has, however, so happened that the late epidemic of smallpox has afforded ample proof, that in overcrowded and low damp houses, and in the neighbourhood of offal-heaps, disease finds its most numerous victims, and there operates with the most deadly effect; many of the places more especially referred to as being filthy, have suffered most from small-pox. In one house where there are about 40 persons living, there have been 13 cases of small-pox, and 12 of these cases were in two families. The windows of the rooms in which these families live, open upon the yard in which are two piggeries, a filthy privy, and a large dung-heap while down the surface of the yard flows the whole of the refuse water from all the families on the premises. There have been 10 cases in the neighbourhood of the yard at the end of the Fore-street before mentioned, where slaughter-houses, piggeries, dung-heaps, stagnant putrid pools, and all the other filthy concomitants are crowded together directly beneath the windows of a row off houses that runs parallel therewith. From the evidence supplied by the medical gentlemen of the town, it is conclusive that three fourths of the cases of small-pox have been in the districts pointed out in this report as in the worst sanitary condition. And there can be no doubt but that physical illness is but one of the many evils consequent upon such a state of things. A clean town is an important step towards having a cleanly and moral population.

     (Signed) R. E. RIDLEY.
         J. HUDSON.
         W. PEARSON.
         THOS. STAINTHORPE."

" BURIAL GROUND REPORT.

" The sub-committee appointed to inquire into the state of the burial-ground belonging to the parish of Hexham, having discharged that duty to the best of their power, now proceed to report to the general committee for procuring the Health of Towns Act, the result of their inquiry.

" 1st. They would state that through the services of one of their number, Mr. H. Walton, they have obtained the admeasurement of the ground now made use of for the interment of the dead, and find that the whole quantity so used amounts to 3,892 square yards.* This they calculate, by allowing each grave to occupy two yards, will admit of 1,946 graves; the population entitled to the use of this ground amounting to 5,528.

" 2d. Your committee have examined the parish registers of burials for the last 21 years, and find that for the first 7 years of that period, viz., from 1830 to 1837, the annual average of deaths or burials has been 121¾; that for the next 7 years, viz., from 1837 to 1844, it has been 118¾; and that for the last 7 years, or from 1844 to 1851, it has been 130. Taking this last average number of burials, and calculating the ground to contain 1,946 graves, they compute that each grave would require to be re-opened at an interval of rather less than 15 years, though in many cases they are no doubt opened much more frequently. It is indeed often very unpleasant to witness the remains of bodies dug out of the ground before they have become sufficiently decomposed. Until three or four years ago the usual depth of the graves was only 3 feet, so that the top of the coffins would be scarcely 2 feet from the surface, a circumstance this which could not but prove injurious to the health of the inhabitants residing in the immediate vicinity of the churchyard. It is, however, satisfactory to know that, at the period mentioned, an alteration was made in this practice, and that now no grave is allowed to be of a less depth than 5 feet. But the sub-committee feel it their duty to notice another, and perhaps greater evil than that just spoken of. This is the custom, still resorted to by some families, of making use of the inside of the church as a place of interment. On inquiry, it has been ascertained that there are nearly fifty families which have availed themselves of this privilege during a comparatively recent period of time.

---

 * With the exception of the Roman Catholic graveyard.

"The difficulty of putting a stop to a practice which, in the present day is almost universally disapproved of, seems to be that any interference on the part of authorities to prevent the burial of the dead in the same graves with their near relatives, is regarded as implying a want of kindness toward the surviving members of the family, whose feelings on such occasions are but too easily wounded. As, however, it is now generally admitted by those who are best able to give an opinion upon the subject, that the interment of the dead within towns, and especially within buildings, is hurtful to the health of the living, and as the exhalations arising from decaying animal bodies cannot but tend to breed pestilential disease, it is earnestly to be hoped that ere long another and more suitable burial-ground than the present one will be provided, and also that the practice of interring the dead within the church will be altogether discontinued.

"Your committee are not aware that they can supply any other information bearing on this subject, they therefore now conclude their report.

(Signed) J. HUDSON.
WM. W. GIBSON.
"*Hexham*, 24th Sept. 1852." HENRY WALTON."

"REPORT of the Visitors of the LODGING HOUSES, HEXHAM, 24th October 1851.

" *Old Burn-lane.*

"No. 1. *Margaret Fairley* alias *Fisher*.—2 rooms; 1 on the ground floor, occupied by the family; 1 room above, approached by a step-ladder, for lodgers, 7 beds; dimensions, 21 feet by 16 feet; wall one foot above the floor, unceiled, thatched roof; number of lodgers, 6 to 12; room lighted by 1 small window."

REMARKS.—Or about 1,680 cube feet of space for 12 persons, being at the rate of 140 cube feet of air to each person, with defective means of ventilation; criminals in gaol have from 800 to 1,000 cube feet of air space each, with the best means of ventilation.

" *Gilligate Wooden Bridge.*

"No. 2. *Robert Riley.*—2 rooms on the ground floor; 1 room 16 × 14 × 6, 4 beds; 1 ditto, 19 × 14 × 6, 2 beds; lodgers 6 to 8; rooms well lighted.

"No. 3. *William Wren.*—2 rooms, first floor; 1 room 13 × 15 × 8, 3 beds; 1 room 15 × 12 × 7, 3 beds; lodgers 6 to 12*; rooms not well ventilated.

REMARKS.—These rooms have about 2,820 cube feet of air space, and are at times crowded by 60 persons, affording only 47 cube feet of air space to each person. The rooms are said to be "not well ventilated." If it would be criminal to poison 60 persons through their food or drink, surely it must be equally

---

* Sometimes 60 in one night.

criminal to poison them through the air they are in such places obliged to breathe.

"No. 4. *Mrs. Bell.*—1 room, third floor, $17\frac{1}{2} \times 13 \times 8$, 3 beds ; 2 windows; lodgers 2 to 6.

"*Barracks.*

"No. 5. *Isabella Boyd.*—2 rooms, first floor ; 1 room $15 \times 11 \times 7$ 9, 5 beds ; 1 room occupied by family ; lodgers 10 to 15 ; 23d October, 15 in one room in one night ; rooms well lighted.

"No. 6. *Mrs. Stewart.*—1 room, $17\frac{1}{2} \times 14$ 8 $\times 9$ 3, 5 beds ; lodgers 8 to 12.

"*Church-way.*

"No. 7. *Peter Doran.*—1 room, second floor, $15\frac{1}{2} \times 11 \times 7$ 3. 3 beds; lodgers 6 to 10 ; family 4.

"*Battle-hill.*

"No. 8. *Wm. Coulson.*—2 rooms ; 1 room on ground floor. $16 \times 11 \times 7$; 2 beds ; 1 room, first floor, $16 \times 11 \times 7$, 4 beds ; lodgers 6 to 8 ; situated in Back-yard.

"No. 9. *Jane Cinnamon.*—2 rooms; 1 room $15 \times 6\frac{1}{2} \times 8$, 2 beds ; lodgers 4 to 8 ; third floor, fronting street.

"*Victoria-place.*

"No. 10. — *Kearney.*—2 rooms on ground floor ; 1 room $17 \times 8 \times 8$ 3, 6 beds; 1 room $17 \times 9\frac{1}{2} \times 8$ 3, for family ; lodgers 12 to 16 ; situated in an open yard.

"No. 11. *Hannah Vannan.*—3 rooms, 2 on ground floor each $14\frac{1}{2} \times 15 \times 7\frac{1}{2}$, 7 beds ; lodgers 5 to 12 ; situated in an open yard.

"*Broad-gates.*

"No. 12. *James Renwick.*—2 rooms, first floor, each 15 4 $\times$ 14 $\times$ 7 6 ; 4 beds in each room; lodgers 8 to 16.

"*Priest-popple.*

"No. 13. *Robert Wilson.*—2 rooms, each $15\frac{1}{2} \times 13\frac{1}{2} \times 7$ 4, 4 beds ; lodgers 4 to 8 ; situated in a yard.

"No. 14. *Ann Fitspatrick.*—3 rooms, 1 on ground floor ; 1 room $14 \times 13 \times 8$, 3 beds ; 1 room $12\frac{1}{2} \times 11 \times 6$ 9, 2 beds ; lodgers 10, family 6 ; situated in a close yard ; not well lighted.

"No. 15. *Joseph Turnbull.*—2 rooms, both on ground floor ; 1 room 13 6 $\times$ 15 3 $\times$ 8, 5 beds, lodgers 8 to 10 ; 1 room occupied by the family, 10 in number ; situated in a close yard, not well ventilated, and has the appearance of dampness.

"No. 16. *Jane Johnston.*—1 room, 11 9 $\times$ 11 4 $\times$ 6 6, 3 beds ; lodgers 5 to 10 ; situated in the same yard as the two preceding.

(Signed)   HENRY WALTON.
                        WILLIAM ELLIS.
" *Hexham, 24th October* 1851."   JOHN OLIVER."

REMARKS.—In these common lodging-houses it is shown that the poor have, in some instances, only about one twentieth the amount of air space it is deemed advisable to afford to the worst criminals when in gaol. This positive poisoning of the atmosphere is not, however, the worst feature of the case, as in these crowded places there is no separation of the sexes, and male and female, youth and age, lie down together in nakedness, as, on account of the heat and vermin, the males regularly strip themselves to the skin, without having the least sense of degradation or shame. Many of the room tenements are crowded to excess, and as there is only one sleeping room, father, mother, and grown up sons and daughters all crowd the same room, and not unfrequently one bed. Incest, under such circumstances, is not uncommon. It is to enable the poor to avoid such a fearful crime, that your Honourable Board is called upon to assist the local gentlemen wishful to apply the Public Health Act.

" REPORT of the Extent, State, &c. of the Public ROADS and STREETS within the Township of HEXHAM, and kept in repair by a highway rate, and under the charge of four surveyors.

" *Extent.*—There are about 5 miles and 978 yards of roads and streets in charge of the above surveyors, 1,700 yards, or nearly 1 mile of which are streets, besides about 2½ miles occupied and partly kept in repair by the trustees of the Newcastle and Carlisle turnpike road, also about 624 yards of the Alemouth turnpike within the township, making in all 8 miles and 702 yards, or nearly 8½ miles of roads and streets within the township.

" *State, &c.*—The roads or highways are macadamised, *i. e.* laid with hard stone, broken small, and are in good repair. The streets are partly penned, partly macadamised, and partly paved with pebble stones. Only *one* street is penned, and the pennstones being freestones are too soft for the weight of carriage coming upon them, hence the street is not in such a good state as might be expected, it not being very long since it was repenned. The paving in other streets needs repairing. The macadamised streets are in good repair.

" *Footpaths.*—The footpaths along the sides of the streets are in a very irregular state. The proprietors of some of the properties have the footpaths in front of such properties flagged, and are generally in good repair; whilst others, the greater proportion (there being no law to compel flagging), have only pebble paving on the footpaths in the fronts of the houses. In one street, viz., the Fore-street, the footpaths have been flagged for many years, yet at the expense of each proprietor, such proprietor flagging to the extent of his property; hence, at present, many portions require to be reflagged. In the same and other streets there is a great annoyance from several of the houses not having spouts at the eaves, consequently, on a rainy day pedestrians are driven into the middle of the street for shelter.

" The Market-place, except a roadway through the centre, is kept in repair by the lord of the manor, he receiving all the fair and market tolls, stall rents, &c.

" The expense of keeping the roads and streets in repair is provided for by an annual rate upon the property of the said township, which rate amounts to about 150*l.* per annum, which is at the rate of 4*d.* in the pound.

"The thoroughfare extending from the foot of Priest-popple to the high end of Hencotes is included in the road under the charge of the trustees of the Hexham turnpike road; yet the footpaths along the fronts of houses in the said thoroughfare, whether flagged or paved, are kept in repair by the several proprietors.

"The trustees of the aforesaid turnpike road are paid 27l. annually out of the above highway rate to assist in keeping the aforesaid township roads in repair.

(Signed)   HENRY WALTON."

"BOUNDARY REPORT.

"The sub-committee of the 'Public Health Act Committee,' appointed to consider the question of boundary, report that they do not find any known boundary so generally suitable for the purposes of the Act as that of the ancient lands of the township of Hexham; and they recommend that it be submitted to the inspector as the boundary within which the Act shall be put in force. It is not improbable that some few landowners may consider this boundary open to objection, but the sub-committee think it better that any such objections should be brought before the inspector by objecting parties, and disposed of by him, than that they should endeavour to anticipate and remove them, in which endeavour it is not likely that they would be entirely successful.

"*Hexham, 13th February* 1852."
     CHAS. HEAD.
     J. GIBSON."

APPENDIX C.

The following particulars, as to the ancient government of the town, were furnished to me by Jasper Gibson, Esq., solicitor. These powers are found, practically, to be nearly useless.

ANCIENT GOVERNMENT.—DIVISION OF TOWN.—OATHS OF LOCAL OFFICERS, &c.

The regality or manor of Hexham, with the members, in the county of Northumberland. } The great court or head court, court leet, and view of frankpledge, together with the court baron of Wentworth, Blackett Beaumont, Esquire, lord of the said manor, holden there the 21st day of April in the year of our Lord 1852; before James Losh, Esquire, learned steward of the same court.

Borough Jury.

Robert Lyon, foreman.
William Alexander.
William Angus.
John Cook.
Joseph Dixon.
Henry Hays.
} Sworn.

James Hill.
John M'Kane.
Thomas Newbigen.
Matthew Ord.
George Pattinson.
Thomas Robinson.
} Sworn.

Affeerors.

Francis Robinson.   George Scott.

Ancient Government.

We present and order, that the persons hereafter named by us the jury appointed, the four-and-twenty of this town, shall tax and lay on cesses in and about the said town, as formerly accustomed, to the best of their skill and judgment; and as often as occasion shall require, consult and advise with the bailiff of all such matters and things as shall be thought fit and convenient for the good of the said town, and that they and every of them shall meet together on warning by the constable, on pain of 6s. 8d. each and every of them offending therein.

Market-street Ward.
Rev. Josh. Hudson.
J. D. Bell.
Richard Mews.
Robert Lyon.
Geo. Pattinson.
Jas. Turnbull.

Priest-popple Ward.
W. W. Gibson.
Ed. Baty.
Geo. Bates.
Thos. Clemitson.
Wm. Hopper.
Thos. Hedley.

Gilligate Ward.
James King.
Thos. Davison.
John Ruddock.
Henry Allen.
Thos. Pratt.
O. Anderson.

Hencotes Ward.
George Porteous.
Jas. Renwick.
Matthew Brown.
Henry Hays.
Wm. Pruddah.
F. Bell.

Scavengers.
John Forster.
Ann May.
Jane M'Kinley.

Sarah Ridley.
Martha Smith.

We present and order, that the scavengers shall, once in every week, sweep the Market-place and pant-gutter, so far as the house occupied by Edward Pruddah; and that they prevent the water from running down Hall-style-head; and also that they sweep the Hall-gate, and that part of the Hall-gate wherein the market is partly holden; and that they shall not lay any dirt under the covered market or shambles on pain of losing their places.

We present and order, that all persons furnishing the fishers with stalls shall cause the fish-guts to be removed out of the Market-place, on pain of 6s. 8d. each and every one offending therein.

We present and order, that no person shall lay any stones, thatch, or rubbish in the burns, Market-place, streets, common lanes, or highways of, in, or about the town of Hexham, on pain of 13s. 4d. each and every one offending therein.

We present and order, that no butcher or other person shall slaughter or dress their sheep, lambs, calves, or swine in the Market-place of Hexham, or throw the blood or entrails of any sheep or other beast therein, on pain of 13s. 4d. each and every one offending therein.

We present and order, that no person shall sharp any bill or knife on the pant-stones, or take both pipes of water at the pant in the Market-place of Hexham aforesaid at one time, unless there be no want of the other, on pain of 13s. 4d. each and every one offending therein.

We present and order, that the inhabitants of the town of Hexham shall not suffer their swine to go at large in the town streets, or out of their yards, on pain of 13s. 4d. each and every one offending therein.

We present and order, that no person shall buy up the butter, cheese, or other provisions coming to Hexham market before one of the clock in the afternoon, on pain of 13s. 4d. each and every one offending therein.

We present and order, that the market-keepers shall try the weights and measures, four times at least in the year, and that one or more of them shall inspect the market every market-day, on pain of 13s. 4d. each and every one offending therein.

We present and order, that the market-keepers shall search every house wherein they suspect corn set up, designed for sale, but not brought into the market; and if they find any, that they seize the same; and that no innkeeper shall suffer any corn to stand in his or her house on the market-day, before one of the clock in the afternoon, on pain of 13s. 4d. each and every one offending therein.

We present and order, that no person shall carry away any soil, dung, or manure off Tyne-green, or lay any bark, lime, rubbish, or other matter or things thereon, on pain of 13s. 4d. each and every one offending therein.

We present and order, that no person shall leave their stalls or or carts standing in the Market-place or streets of Hexham in the night-time, on pain of 13s. 4d. each and every one offending therein.

We appoint John Baty inspector or examiner of weights and measures in and for the township of Hexham, in the regality or manor of Hexham, under the provisions of the Act of 5 & 6 W. 4. cap. 63.

(POWERS ENFORCED).

We present William Robb for contempt of court, and amerce him in the sum of 2s. 6d.

We present Robert Headley for digging out and laying up a quantity of soil on Tyne-green, 10s., and if not levelled down as found, 15s. A further sum, for the first offence we amerce him in the sum of 10s., and for the second, 15s.

We present Wilkinson Hunter for leading away a quantity of soil from Tyne-green, and amerce him in the sum of 20s.

Robert Lyon.

John M'Kane.	George Pattinson.
Henry Hays.	James Hill.
Thomas Robinson.	William Alexander.
William Angus.	John Cooke.
Joseph Dixon.	Thomas Newbiggen.
Matthew Ord.	

We affeer the above presentment,
GEORGE SCOTT.
FRANCIS ROBINSON.

Constable's Oath.

You shall well and truly serve our sovereign Lord the King, and the lord and lady of this leet, in the office of a constable in and for the   of   in all the duties of a constable, unconnected with the preservation of the peace, or the execution of an Act passed in the sixth year of the reign of Her Majesty, intituled "An Act for the appointment and payment of Parish Constables," until you shall be thereof discharged by due course of law. You shall well and truly do and execute all things belonging to the said office, according to the best of your skill and knowledge. So help you, &c.

The Oath of the Market-keepers, Appraisers, and Sealers.

You shall swear that you will well and truly serve His Majesty, and the lord and lady of this manor, in the office of market-keepers, appraisers, and sealers of this town, for this year to come, or till others be chosen and sworn in your rooms. You shall duly and truly, from time to time, see that the assize of bread be observed and kept according to the laws and statutes of this kingdom after an assize is set, and that the bread brought to be sold be duly weighed before the same shall be exposed to sale in the market; and that the same do continue such weight according to the prices of wheat as by the statute in that behalf is provided. You shall also inquire of all conspiracies of butchers and bakers not to sell victuals but at certain prices, and of all other faults committed by butchers, bakers, brewers, and tipplers, or any of them, and of all that use any false weights, the greater to buy with and the lesser to sell with; and you shall diligently assay and try the weights and measures of all merchants, pedlars, petty tradesmen, and chapmen, and all corn measures, and see they be agreeable to the Winchester measure by statute appointed to buy and sell with. You are, to the utmost of your power, to endeavour that no corn be sold in the market by any other measure than the Winchester measure, nor that any corn be measured by any other measure than by such as is marked by the towns burn. You are as well to make known and present all offences done by any person or persons against the tenor of this your oath, as also all unlawful forestalling, ingrossing, and regrating, that punishment according to the law may be inflicted on all offenders. You are to present those that sell corn in the market before the cornbell rings. You shall also well and truly, to the best of your judgment, appraise and value all such goods and chattels as shall either be attached or taken upon executions by any of the officers of the court of this manor; as also all such damages as shall be occasioned by reason of any trespasses done or committed in any of the corn fields, meadows, or pastures of this town. You shall be ready and attend upon the lord and lady of this manor, steward of this court, and bailiff of this manor, at all time and times when you shall be thereunto by them or any of them required so to do; and you shall well and truly, according to the best of your cunning and power, do all things incident and appertaining to the said offices of market-keepers, appraisers, and sealers, till others be chosen in your rooms.

So help, &c.

The Common-keeper's Oath.

You shall swear, that you, to the best of your power, shall see and have regard that the hather and grass growing on the commons belonging to this town be kept and preserved for the benefit and use of the inhabitants thereof. You shall impound or cause to be impounded the goods and cattle of such persons as you shall know trespass on the said commons, and have no right on the same. You shall also present all such person or persons as you shall see burn or destroy the said hather, and all those who pull any of the same, having no right thereto, without the leave and consent of the bailiff and inhabitants of this town first had and obtained; and also shall do and perform everything belonging to the execution of your said office.

So help, &c.

The Surveyor's Oath.

You shall swear, that you shall well and truly serve His Majesty, and the lord and lady of this manor, in the office of a surveyor for this town, for one whole year and a day, or till the next head court to be held for this manor, or till another be chosen and sworn in your place; and you shall use all diligence and care you can to cause all bridges, causeways, and highways to be duly amended and repaired. You shall diligently inquire and true presentment make of all common nuisances in and about this town or the precincts thereof, and of the person or persons that cause or procure such common nuisances. You shall be ready and attend upon the lord and lady of this manor, steward of this court, and bailiff of this manor, at all time and times when you shall be thereunto by them or any of them required so to do; and you shall in all things well and truly, to the best of your cunning and power, do all things incident and appertaining to the said office of a surveyor, till another be chosen and sworn in your stead.

So help, &c.

The Oath of the Four-and-twenty.

You shall swear, that you and every of you, within the precincts of your several divisions and wards in Hexham for which you are chosen, shall well, truly, indifferently, and conscientiously tax and assess such persons within the said wards as shall from time to time and as need may require be assessed or taxed, touching the raising of monies for this town's use; and shall, to the best of your understandings, cunnings, and powers, be careful and provident in all things which may conduce to the benefit, welfare, and common-wealth of this said town and the inhabitants thereof, and the order of the court to that end made.

So help, &c.

The Ale-taster's Oath.

You shall swear, that you shall well and truly serve our sovereign Lord the King, and the lord and lady of this manor, in the office of an ale-taster or assisor of this town, for this year to come.

You shall have diligent care during the time of your office, to all the brewers and tipplers within your office, that they and every of them do make good and wholesome beer and ale for man's body, and that the same be not sold before it be assayed by you, and then to be sold agreeable to the prices limited and appointed by the justices of the peace; and all faults committed or done by the brewers or tipplers or by any of them you shall make known and present the same at this court, where due and condign punishment may be inflicted upon them for such their offences accordingly, and in every other thing you shall well and truly behave yourself for this year to come.

So help, &c.

LONDON:
Printed by GEORGE E. EYRE and WILLIAM SPOTTISWOODE,
Printers to the Queen's most Excellent Majesty.
For Her Majesty's Stationery Office.

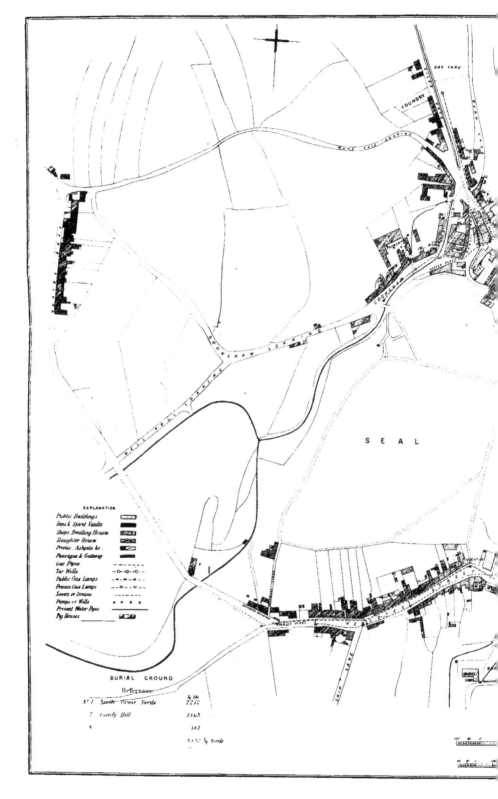

PART OF
THE TOWN OF HEXHAM
(near the Abbey Church)
DRAWN TO ENLARGED SCALE FROM SURVEY MADE FOR THE PURPOSES OF
THE PUBLIC HEALTH ACT, 1848.

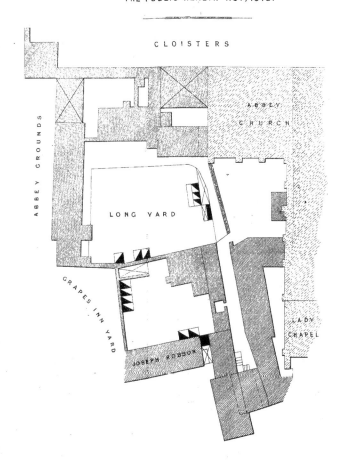

EXPLANATION.

Public Buildings
Shops, Dwelling Houses
Slaughter Houses
Privies, Ashpits, &c.
Pig Houses
Passages & Gateway
Area of the Burial Ground 3892 Square Yards.

NOTE.

This Plan shews the crowded state of the dwelling houses, the vicinity of pigsties, slaughter-houses and privies.

Scale of Feet

Lightning Source UK Ltd.
Milton Keynes UK
UKHW02f1600110418
320862UK00006B/749/P